Genetics, Biotechnology and Breeding of Maize and Sorghum

Genetics, Biotechnology and Breeding of Maize and Sorghum

Edited by

A.S. Tsaftaris
Aristotelian University of Thessaloniki, Greece

THE ROYAL
SOCIETY OF
CHEMISTRY
Information
Services

The Proceedings of the XVIIth Conference on Genetics, Biotechnology and Breeding of Maize and Sorghum held at the Aristotelian University of Thessaloniki, Greece on 20–25 October 1996.

Special Publication No. 209

ISBN 0-85404-762-X

A catalogue record for this book is available from the British Library

Published by The Royal Society of Chemistry,
Thomas Graham House, Science Park, Milton Road,
Cambridge CB4 4WF, UK

Printed by Bookcraft (Bath) Ltd.

Preface

Maize was involved in the rediscovery of Mendelism and continues to play an important role in advancing in our knowledge of plant genetics, breeding and now in plant molecular biology and biotechnology. In addition, maize, and to a lesser degree, sorghum are of the most important crops. The use of hybrid seed, the designing of breeding schemes for maximing F_1 heterosis for yield and many of the recent developments of biotechnology are well exemplified in maize.

This volume contains a collection of papers devoted to all aspects of genetics, biotechnology and breeding of maize. It is an indication of our present state of knowledge of this very important and broad field of information on recent studies in genome organization and evolution of the cereal genome, maize and sorghum transformation and transgenic genotypes developed and used, QTL analysis and use in maize and breeding for stress tolerance and low input demands. Particular emphasis in this volume is given on the use of maize as a model and good representative species for all aspects of genetics, molecular biology, biotechnology and breeding of cereals in general. This book describes approaches and methods of maize breeding in ways which increase our potential of combining conventional and modern techniques.

The papers of this volume are written presentations made in an International Conference entitled: "Genetics, Biotechnology and Breeding of Maize and Sorghum". The conference was organized in Thessaloniki, Greece, during October 20-25, 1996 and it was the 17th Conference of the Maize and Sorghum Scection of Eucarpia.

I would like to thank the members of the Organizing Committee for their effort to organize this Conference and the Sponsporships of HELLAS EED, Co. The financial support of EUCARPIA, European Union, University of Thessaloniki, National Agricultural Research Foundation of Greece and a number of private companies is also acknowledged.

A.S. TSAFTARIS

Contents

Session V - Breeding Maize for Stress Tolerance and Low Input Demands

SESSION I
THE CEREAL GENOME: FROM RICE TO MAIZE

Comparative Genome Analysis in the Grass Family

Katrien M. Devos and Mike D. Gale

JOHN INNES CENTRE, NORWICH RESEARCH PARK, COLNEY, NORWICH NR4 7UH, UK

1. INTRODUCTION

Plant genetics has entered a new era. The development of restriction fragment length polymorphisms (RFLPs) in the mid 1980s provided researchers with a tool to construct detailed genetic maps. Within a time span of 5 years, the genomes of most major crop plants were mapped, agronomically important genes were being tagged and the hunt for quantitative trait loci (QTLs) had started. Genetic analyses progressed independently in individual species, although similar traits were often targetted. This is not surprising, as it appeared unlikely that the genomes of species such as rice, maize and wheat would be conserved after more than 60 millian years of separate evolution. Today, we know that not only gene content but also gene orders are highly conserved among species within the grass family. This paper will take you through the developments of comparative genetics, from the comparison at the map level of genomes within polyploid species to the comparison at the microlevel of genomes of species belonging to different tribes within the family of the *Poaceae*.

2. COMPARATIVE GENETIC MAPS

The construction of genetic maps of bread wheat (*Triticum aestivum*; $2n=6x=42$, AABBDD) lagged behind that of other crops due to its large genome size ($1C= 1.7 \times 10^{10}$ bp), its high content of repetitive DNA and its hexaploid nature. Hybridization of single copy cDNA probes to digested genomic wheat DNA identified three fragments located on each of the three wheat genomes (Fig.1). Conservation of gene content across the A, B and D genomes was no new concept as triplication of agronomic traits and isozymes had already been observed (for catalogue see McIntosh et al. [1]). The genetic maps of hexaploid wheat, however, provided the first evidence for an extensive conservation of colinearity among the A, B, and D genomes. This colinearity was interrupted only by a complex set of evolutionary translocations involving chromosomes 4A, 5A and 7B [2] and, probably, a reciprocal translocation involving chromosome arms 2BS and 6BS [3]. Colinearity, however, was maintained within each of the translocated segments.

Using common sets of probes, the comparative mapping effort was then extended to include different species within the Triticeae tribe. A comparison of the genomes of *T. monococcum*, *T. tauschii*, barley and wheat revealed nearly complete colinearity [3-6]. Other

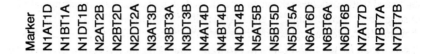

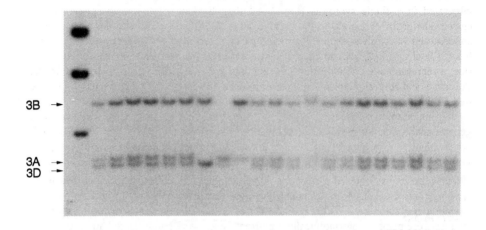

Figure 1 *Hybridization a wheat clone,PSR910, to DNA of 21 nullisomic-tetrasomic lines digested with HindIII identifies fragments on chromosomes 3A, 3B and 3D*

Triticeae species, such as rye and *Aegilops umbellulata* displayed a minimum of 8 and 11 rearrangements, respectively, relative to wheat [7,8]. Once more, colinearity was conserved within each of the translocated segments. The extensive rearrangements that characterise the divergence of some species, *e.g* rye and *Ae. umbellulata*, and not others, *e.g.* barley raises the question as to why certain species accumulate and fix structural changes at a faster rate than others. To date, there is no answer. It is clear, however, that the extent of chromosomal rearrangements is not necessarily a measure for evolutionary age.

Comparative work was also carried out within the Andropogonodae, which includes the crop species maize, sorghum and sugar cane. Following the lead of Hulbert et al. [9], several groups established the relationship between the sorghum and maize genomes [10-15]. The most complete information to date is provided by Dufour et al. [14]. Despite the three- to four-fold larger DNA content of maize compared with sorghum, gene orders appeared to be largely conserved and only a limited number of rearrangements were identified. Extensive colinearity was also observed between sorghum and sugar cane. Modern sugar cane varieties have originated through introgression of part of the genome of the wild species *Saccharum spontaneum* (x=8, $2n$=40 to 128) into *S. officinarum* (x=10, $2n$=80). Introgressions are mostly complete chromosomes but some intergenomic translocations have been observed [16]. The comparative maps proved invaluable in unravelling the intricate structure of the sugar cane genome.

The high degree of conservation in gene orders between the genomes of species within a tribe urged geneticists to extend the comparative research to the family level. The first comparisons were carried out between the genomes of rice (n=12; 1C=0.4pg) and maize (n=10; 1C=2.8pg)[17], and rice and wheat (n=21; 1C=17pg)[18]. Following their integration it emerged that the genomes of maize and wheat could be described as an assembly of 19 rice chromosome segments [19]. Placing these 19 rice segments on a circle, one could arrange the maize and wheat genomes relative to rice. Based on published data on comparative studies of sorghum [9-15], sugar cane [14,20] and foxtail millet [21], Moore et al. [22] were able to show that their model could be applied to a range of grass species. To date, the genomes of seven species, rice, foxtail millet (*Setaria italica*), sugar cane, sorghum, maize, wheat and oats (data from Van Deynze et al. [23]), belonging to three different subfamilies, have been integrated in a single synthesis (Fig.2). Although many of the rearrangements that distinguish the different grass genomes have taken place during or after speciation, several are observed that typify a taxonomic group. In fact, based on common arrangements of rice linkage blocks in Fig.2, one can clearly identify three distinct genome patterns, provided by rice, a representative of the Bambusoideae, oats and the Triticea crops belonging to the Pooideae, and several members of the Panicoideae. The latter group is discussed in more detail below.

It is important to note that rice is used as the base simply because it is the smallest cereal genome analysed in detail,with the densest genetic maps and for which most genomic tools are available. The organisation of Fig.2 has no bearing on the ancestry of the grass genomes. Inferences can, of course be made concerning which chromosomal arrangements are the more primitive.

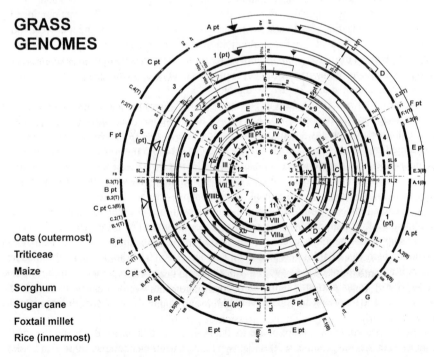

Figure 2 *Aligned maps of rice, foxtail millet, sugar cane, sorghum, maize, the Triticeae crops and oats. All maps are aligned relative to rice. Ends of maps are shown by black arrows. Where known, centromeres locations are indicated. Double ended arrows show inversions relative to rice, open arrows show segments that require transposition to return to linear chromosome maps. Hatched areas indicate regions of uncertainty where the cross-mapping data do not allow definite alignment with rice. An intergenomic duplication between rice 11 and 12 is considered as part of rice 12 only in this analysis.*

3. THE PANICOIDEAE SUBFAMILY

One of the early observations that resulted from comparative genome research was the fact that the maize genome could be aligned with two complete rice genomes, with maize chromosomes 1, 2, 3, 4 and 6 forming one rice complement, and maize chromosomes 5, 7, 8, 9 and 10 a second. The presence of extensive duplications within the maize genome [24] had previously led to speculations about a tetraploid origin. The ancient tetraploid status of maize was now beyond doubt. The fact that most of the larger maize chromosomes make up one genome and the smaller chromosomes make up the other, may relate to differences retained from the time of allotetraploidization. However, the chromosome size differences could also be fortuitous. Sequence analysis of duplicated genes (Doebley, pers. comm.) has suggested that maize is a segmental allotetraploid formed through hybridization of two $2n=5$ species that recombined, and subsequent allopolyploidization with pairing being restricted to homologues rather than homoeologues. More evidence is yet required to support this theory.

The close taxonomic relationship between maize and sorghum, both belonging to the supertribe Andropogonodae and with a haploid chromosome number of 10, raises the question of whether sorghum may also be a tetraploid. Dufour[14] observed that about 23 % of the RFLP probes identified more than one hybridizing fragment in Southern analysis, 11 % of which could be mapped to duplicated loci. A similar figure (8 %) was obtained by Chittenden[25]. Comparing these values with maize, where some 70% of the loci were found to be duplicated [10,26], evidence for a tetraploid origin of sorghum is not strong. Certainly, comparative maps of maize and sorghum show single sorghum linkage groups relating to, usually, two regions of the maize genome, as indicated in Fig.2. For these reasons, sorghum is currently maintained as a $2x=2n=20$ diploid.

Foxtail millet is a staple crop in Northern China. Taxonomically, the crop is placed within the supertribe Panicodae, which is part of the Panicoideae subfamily. Genetic maps of foxtail millet, which has a genome size comparable to rice (450 Mb; $2n=18$) were first constructed by Wang et al. [27] in an intraspecific cross between two *S. italica* accessions. A comparison with a genetic map constructed in an interspecific cross between *S. italica* and its wild relative *S. viridis* demonstrated that both maps were highly similar in marker order and genetic length [27]. Alignment of these maps with the rice genome demonstrated that five foxtail millet chromosomes were colinear with one rice chromosome each, while the remaining four had synteny with two rice chromosomes. Also interesting is the close relationship at the genome level between foxtail millet with nine chromosomes (*n*) and sorghum which has a haploid chromosome number of 10. The difference in chromosome number is accounted for by the synteny of foxtail millet chromosome III with sorghum E and I. Elsewhere we can currently detect only one inversion, in sorghum chromosome D, and one translocation involving foxtail millet chromosomes III and VII, which differentiate the two species.

A comparison of the structure of the foxtail millet, sugar cane, sorghum and two maize genomes with the rice genome, reveals that two arrangements characterise these Panicoideae species; the intercallation of rice chromosome 9 into rice 7, and the rice 10/rice 3 arrangement (Fig.2). Rearrangements common to groups of species predate their divergence and can be used for taxonomic purposes. Others, such as the rice 5/rice 12 arrangement forming foxtail millet chromosome III are specific to the species and arose during or after their divergence from members of the taxonomic group.

4. APPLICATIONS OF COMPARATIVE RESEARCH

4.1. Extended Marker Pools

One of the obvious advantages of comparative research is the availability of increased marker pools. Genes are highly conserved at the DNA level, and will readily cross-hybridize across species within a family. Furthermore, as gene orders are also maintained, markers can be preselected to lie within the target region for trait mapping or map saturation, or to give good genome coverage for directed mapping of 'new' crop species.

Integrated genetic maps provide additional markers, but the numerous mapping populations themselves, generated in various crops, can also be exploited. Using isogenic lines or a bulked segregant approach, one can identify molecular markers that are linked to agronomically important but, as yet, unmapped traits. The chromosomal location of these traits can then be established by mapping the linked marker(s) in existing mapping populations of that crop, provided the marker(s) exhibit variation. This may be a problem in crops with inherent low levels of polymorphism such as wheat and sorghum. The use of heterologous mapping populations generated from highly polymorphic parents, and with known genome structure in relation to the grass map may provide an elegant way to determine, through extrapolation, the location of traits.

4.2. Centromere Positions

Positioning of centromeres on the genetic maps is achieved by mapping RFLP markers to chromosome arms using appropriate genetic stocks and has, so far, been limited to wheat, maize and rice. As these species have 7, 10 and 12 chromosomes respectively, centromere positions cannot all be conserved. Map comparisons showed that the centromeres of maize chromosomes 1, 2, 3, 5, 7, 8, 9 and 10 correspond to those of rice chromosomes 8, 4, 1, 2, 7, 1, 6 and 4 respectively. The centromeres of maize chromosomes 4 and 6 are homoeologous to the centromeres of either rice 2 or 11, and 5 or 6 respectively, but further data are needed to unravel the precise correspondence. Colinearity between rice and wheat centromeres was also established for five of seven wheat chromosomes. Since colinearity appears to extend to centromere positions, one can use comparative genetics to predict centromere locations in crops for which no ditelosomic lines or other appropriate stocks are available. Using this approach, putative centromere positions were established for foxtail millet chromosomes I, II, IV, V, VII, VIII and IX [21]. Particularly on chromosomes I, IV, V, VI and IX, the putative centromere regions coincide with regions of severe marker clustering. Reduced recombination around the centromeres was first observed in Triticeae species [28], and supports the hypothesis that the identified regions contain functional centromeres.

4.3. Gene Isolation

In addition to determining the chromosomal and map locations of traits, it is often desirable to isolate the gene's underlying traits to study their mode of action. Gene isolation through chromosome walking (or landing) is feasible in species with small to moderate genome sizes, such as rice and sorghum, but has, unfortunately, little chance of success in large genome species such as the Triticeae cereals. Main problems are the instability of large insert YAC and BAC clones due to the presence of repetitive DNA sequences, and the vast number of clones that are needed even to achieve only a two times genome coverage. These problems could be overcome by using a small genome size species such as rice as an intermediate to isolate, for example, a barley gene, provided that the conservation of gene orders at the map level extends to the sub-cM level. Dunford et al.[29] demonstrated that the order of markers, contained on the same YAC and thus separated by less than 1 Mb in rice, was maintained in barley. Similar comparative studies demonstrated colinearity at the microlevel between regions of barley chromosome 7H with rice 6[30] and wheat chromosome 5 with rice 9[31]. Despite this general observation of colinearity at the sub-cM level, both Kilian's[30] and Foote's[31] studies found small duplications that disrupted gene orders between rice and their target crop. Chen et al.[32] demonstrated a duplication of the *a1* gene in sorghum, with only one copy present in rice and maize. The *Sh2-a1* region, however, exhibited extensive colinarity in all three species, but spanned 140 kb in maize, and only 19 kb in rice and sorghum. Sequence homology between orthologous genes amounted to at least 80 %, while most of the intergenic regions were repetitive and failed to cross-hybridize across species[32].

In practical terms this means that we can now use markers, tightly linked to the gene of interest in the target crop, to undertake a chromosome walk and isolate the gene in a colinear region of, for example, the rice genome. The rice gene can, in turn, be employed to pull out the homologue in the target crop. This approach is being used at JIC, Norwich and WSU, Pullman, to search for the wheat *Ph1* gene controlling homologous pairing, and *Rpg1*, a barley rust resistance gene, respectively.

5. CONCLUSIONS

Comparative genome analysis has demonstrated a high conservation of gene content and gene orders between species within the grass family. These studies allowed the integration of the genomes of seven species, rice, foxtail millet, sugar cane, sorghum, maize, the Triticeae cereals and oats, belonging to three subfamilies, in a single synthesis, and work is underway to extend the grass comparative map with pearl millet (*Pennisetum glaucum*), ryegrass (*Lolium perenne*) and finger millet (*Eleusine coracana*). Rearrangements, both at the map and megabase level have taken place, but 60 million years of evolution has left large chromosome regions apparently untouched. This extensive colinearity can now be exploited to map traits more efficiently, to isolate genes in crops irrespective of their genome size, and to integrate the knowledge on biochemical and physiological pathways in all grasses, thereby providing a quantum leap forward in our understanding of *Poaceae* species.

References

1. R.A. McIntosh, 'Proc 7th Int Wheat Genet Symp',(Eds.)T.E. Miller and R.M.D. Koebner, IPSR,Cambridge Laboratory, Cambridge, 1988, p. 1225.
2. K.M. Devos, J. Dubcovsky, J. Dvorák, C.N. Chinoy and M.D. Gale, *Theor.Appl.Genet.*,1995, **91**, 282.
3. K.M. Devos, T. Millan and M.D. Gale, *Theor.Appl.Genet.*,1993, **85**, 784.
4. J. Dubcovsky, M.C. Luo, G.Y. Zhong, R. Bransteitter, A. Desai, A. Kilian, A. Kleinhofs and J. Dvorák, *Genetics*,1996, **143**, 983.
5. D.M. Namuth, N.L.V. Lapitan, K.S. Gill and B.S. Gill, *Theor.Appl.Genet.*,1994, **89**, 865.
6. A.E. Van Deynze, J. Dubcovsky, K.S. Gill, J.C. Nelson, M.E. Sorrells, J. Dvorák, B.S. Gill, E.S. Lagudah, S.R. McCouch and R. Appels, *Genome*,1995, **38**, 45.

7. K.M. Devos, M.D. Atkinson, C.N. Chinoy, R.L. Harcourt, R.M.D. Koebner, C.J. Liu, P. Masojc, D.X. Xie and M.D. Gale, *Theor.Appl.Genet.*,1993, **85**, 673.

8. H. Zhang, K.M. Devos, J. Jia and M.D. Gale, *Theor.Appl.Genet.*,1997, (In Press)

9. S.H. Hulbert, T.E. Richter, J.D. Axtell and J.L. Bennetzen, *Proc.Nat.Acad.Sci.*,1990, **87**, 4251.

10. R. Whitkus, J. Doebley and M. Lee, *Genetics*,1992, **132**, 1119.

11. A. Melake Berhan, S.H. Hulbert, L.G. Butler and J.L. Bennetzen, *Theor.Appl.Genet.*,1993, **86**, 598.

12. M.G. Pereira, M. Lee, P. Bramel-Cox, W. Woodman, J. Doebley and R. Whitkus, *Genome*,1994, **37**, 236.

13. A.H. Paterson, Y-R. Lin, Z. Li, K.F. Schertz, J.F. Doebley, S.R.M. Pinson, S-C. Liu, J.W. Stansel and J.E. Irvine, *Science*,1995, **269**, 1714.

14. P. Dufour, *PhD Thesis,Université de Paris Sud,France*,1996.

15. P. Dufour, L. Grivet, A. D'Hont, M. Deu, G. Trouche, J.C. Glaszmann and P. Hamon, *Theor.Appl.Genet.*,1996, **92**, 1024.

16. A. D'Hont, L. Grivet, P. Feldmann, P.S. Rao, N. Berding and J.C. Glaszmann, *Mol.& Gen.Genet.*,1996, **250**, 405.

17. S. Ahn and S.D. Tanksley, *Proc.Nat.Acad.Sci.*,1993, **90**, 7980.

18. R. D'Ovidio, O.A. Tanzarella and E. Porceddu, *Plant Mol.Biol.*,1990, **15**, 169.

19. G. Moore, T. Foote, T. Helentjaris, K.M. Devos, N. Kurata and M.D. Gale, *Trends Genet.*,1995, **11**, 81.

20. L. Grivet, A. D'Hont, P. Dufour, P. Hamon, D. Roques and J.C. Glaszmann, *Heredity*,1994, **73**, 500.

21. K.M. Devos, Z.M. Wang, J. Beales, T. Sasaki and M.D. Gale, *Theor.Appl.Genet.*,1997, (In Press)

22. G. Moore, K.M. Devos, Z. Wang and M.D. Gale, *Current Biol*,1995, **5**, 737.

23. A.E. Van Deynze, J.C. Nelson, L.S. O'Donoughue, S. Ahn, W. Siripoonwiwat, S.E. Harrington, E.S. Yglesias, D.P. Braga, S.R. McCouch and M.E. Sorrells, *Mol.& Gen.Genet.*,1995, **249**, 349.

24. T. Helentjaris, *Maize Genetics Cooperation Newsletter*,1995, **69**, 67.

25. L.M. Chittenden, K.F. Schertz, Y.R. Lin, R.A. Wing and A.H. Paterson, *Theor.Appl.Genet.*,1994, **87**, 925.

26. J. Dowty and T. Helentjaris, *Maize Genetics Cooperation Newsletter*,1992, **66**, 106.

27. Z.M. Wang, K.M. Devos, C.J. Liu, J.Y. Xiang, R.Q. Wang and M.D. Gale, *Theor.Appl.Genet.*,1997, (In Press)

28. S. Chao, P.J. Sharp, A.J. Worland, E.J. Warham, R.M.D. Koebner and M.D. Gale, *Theor.Appl.Genet.*,1989, **78**, 495.

29. R.P. Dunford, N. Kurata, D.A. Laurie, T.A. Money, Y. Minobe and G. Moore, *Nucleic Acids Res.*,1995, **23**, 2724.

30. A. Kilian, D.A. Kudrna, A. Kleinhofs, M. Yano, N. Kurata, B. Steffenson and T. Sasaki, *Nucleic Acids Res.*,1995, **23**, 2729.

31. T. Foote, M. Roberts, N. Kurata, T. Sasaki and G. Moore, *Theor.Appl.Genet.*,1997, (In Press)

32. M. Chen, P. SanMiguel, A.C. de Oliveira, S.S. Woo, H. Zhang, R.A. Wing and J.L. Bennetzen, *Proc.Nat.Acad.Sci.*,1997, (In Press)

Maize Genome: Factors Contributing to its Reorganization

Peter A. Peterson

DEPARTMENT OF AGRONOMY, IOWA STATE UNIVERSITY, AMES, IA 50011, USA

1 INTRODUCTION

That the maize genome is highly polymorphic is self evident from the numerous studies illustrated by investigations of various domestic lines and races of maize[1,2] and the examination of duplicated sequences prevalent in the genome.[3] How then has this extensive change and reorganization of the maize genome originated? Could transposons be a strong force in the induction of this diversity?

Variegation is a prevalent feature among plants. This was early observed by Darwin, Lecoq, and DeVries.[4] In the development of the early hypotheses regarding this variegation, it was proposed that these somatic and germinal changes were reversions from an ancestral form or atavism.[5]

Another early consideration for this variegation was that they were "sick-genes". This theory proposed by Correns[6] whereby the number of atoms (maximum or minimum of the gene) determines the difference. This was to accommodate the colorful patterns which were extensively diverse in a graded series of coloration in *Capsella*. Also early in the consideration of the gene theory was Emerson's[7] postulate that this instability was related to changes in genes. This was an early and more accurate hypothesis regarding these changes; however, the question remained,what was the nature of the gene change that gave rise to this variegation? Such was the level of understanding until the McClintock experiments.

It wasn't until the McClintock[8,9] bridge-breakage-fusion experiments,leading to the current understanding of the meaning of variegated phenomena,that led to the current and well-documented transposon concept that explains variegation. From these experiments, there originated a large number of unstable alleles that underwent an intensive genetic analysis and eventually resulted in the current recognition of the mechanism leading to variegation.[10A] That the genomes have mobile elements has now been amply documented over the last four decades.[10B] But the prevalent question though largely inferred is whether the variegation is contributing to the vast polymorphism and duplication that McClintock[11] expounded in her address during the Nobel-Prize ceremony in Stockholm. Evidence relating to this question is considered.

1.1 Gene Polymorphism and Duplicate Loci

1.1.1 Polymorphism. The extensive examination of the races of maize illustrate that the isozyme variation uncovers four to five alleles at each locus.[12] This is evident among many enzymes that are tested among these races[13] which is a strong indication that extensive polymorphism exists. A shortcoming in this hypothesis is that the nature of the specific alteration has not been elucidated; however, from other studies it is quite evident that nucleotide sequences have been altered. Further, the nature of the nucleotide sequences and their origin is not self evident, but again other studies would suggest that transposons have been a contributory cause to the nature of the change.

1.1.2 Duplicate loci. As each gene has been uncovered in maize studies, a second locus is apparent in Southern blotting experiments. When this second locus is uncovered, it is generally found to be a duplicate of the original locus and shares close to 95 − 99% base sequence homology in the codogenic region. Among the genes to show duplication in the genome are the chalcone synthase genes [*C2* and *Whp*[14]]. Another case of duplication includes the *C1* and *Pl* loci, also showing considerable sequence homology but functional differentiation by Cone et al.[15] The *R* and *B* loci also are functionally distinct but the sequence homology is considerable between the *R* and *B* genes.[16] The expression of these genes is illustrated in Figure 1 (A to D).

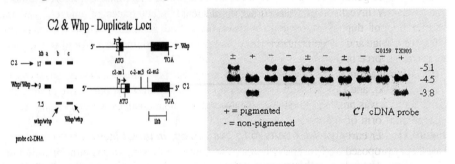

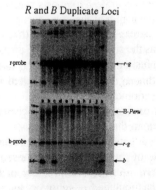

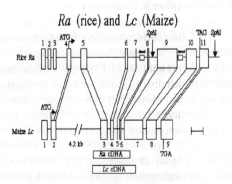

Figure 1. *Southerns illustrating the duplicate nature of several genes*

A.   The *C2* and *Whp* gene arrangements are shown. The *C2* probe lights up both the *C2* and *Whp* bands in the Southern.[14]

B.   The *C1* probe lights up the two *Pl* alleles at 5.1 and 3.8 and the *C1* allele is at the 4.5 band.[15] In this population the *Pl* alleles are polymorphic and *C1* is not. And the probe used was from the coding region fragment from *C1*.

C.   *R* and *B* duplication. These are two Southern blots containing lines segregating different *b* alleles but with the same *r* allele. Top blot, hybridized with an *r* probe and the bottom plot hybridized with a *b* probe. Both the *r* and *b* probes light up the *r* and *b* bands.[16] (Courtesy of V. Chandler)

D.   *Ra* rice and *Lc* maize illustrate differences in the gene structure but show a colinearity in gene order. The two *stoaway* retrotransposons are present in the introns of the rice gene.[17]

In a cross-genera comparison (rice and maize), the waxy probe shows the two genes are colinear in exon order (*Ra* rice and *Lc* maize). It is clear that these two genes are related with introns expanded in the *Ra* gene,[17] (See figure 1E)

The *C1* locus shows two unexpected inserts (6 kb and 250 bp) in the flanking sequence. These are in the noncoding region and have no effect on functionally.

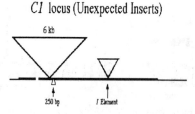

Figure 1E. *Two unexpected inserts (6 kb and 250 bp) in the C1 locus*

Both the polymorphism examples and the duplicate loci examples illustrate the extensive changes that have occurred among genes.

1.2 Stress-induced Volatility

1.2.1 Genome changes. There are a number of stress-induced features that have been related to transposon movement. One of the first was the striking number of unstable alleles that arose out of McClintock's bridge-breakage-fusion cycle events.[8,10] A causal relationship here is only indirect; however, a test of this by Doerschug[18] resulted in a movement of the *Dt* transposon from chromosome 9 to another site.

Wessler has reviewed the numerous cases of stress-induced transposon changes. Certainly tissue culture regimes have resulted in changes in the genome.[19]

1.2.2 Segment Movement. A striking induced genomic segment movement following a bridge-breakage or loss event was the case of the 3-component transposon element system uncovered by a mutant at the *C2* locus. Among the progeny of this breakage event, an *En* deficient related transposon that included a genomic segment unrelated to this transposon moved into the *c2* locus. Such an event was uncovered following the chromosome breakage event which can indirectly be ascribed to the loss-event induced movement of this unusual transposon into a gene locus.[20]

1.3 Gene Changes Following Transposon Excision

During the insertion process of a transposon, the result is a duplication of some sequences. These are called Target Site Duplications (TSD). Each of the transposon families have their "signature" with respect to the number of nucleotides that are added in the TSD. For example with the *En/Spm* system, three nucleotides are duplicated resulting in three of the original host and three of the duplication (See figure 2). With the *Ac-Ds* family, the additional nucleotides added to the host genome include eight, making this total 16. Other families have additional numbers of nucleotides in the origin of the TSD.

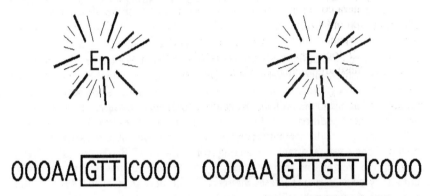

Figure 2. *A diagramatic illustration of the origin of a target site duplication. En/Spm generates a 3-base pair addtion (TSD) upon insertion*

The nucleotide base pairs of the *Ac/Ds* insertion represents a "substrate" for change. This arises from the inaccuracy in the excision process. During transposon excisions, some of the nuclear strands are single stranded and therefore are subject to DNA endonucleose degradation. Further, there are errors in the copying process.[21] As a consequence, a large percentage of the excision events leads to sequences with new codons which would, of course, result in changed proteins. This has originally been cited as a basis for the evolutionary process following transposon inserts.[22] In an early study of this event of the excision of the *En/Spm* system, Schwarz-Sommer found 16 of 18 recovered sequences from *En*-induced excisions at the *wx* locus to be altered. Of the 16, six were frame-shift and eight were with an added leucine, one was an added serine, and two had a deficiency in lysine and phenylalanine. There have been numerous examples of changes at the waxy locus following the excision of *Ac* leading to added amino acids in the new sequence.[23] Changes in the promoter of *Antirrhinum* with the *TAM* transposon have also been found to increase promoter activity.[24]

1.4 Longevity of a Transposon on a Chromosome

In pursuing an excision from the *alm(Au)* allele, an *En* excised from the locus transposed to a nearby site.[25] This particular *En* and the numerous progeny derivatives were followed for four generations and in the progeny of crosses *En* remained on the chromosome 50% of the time. In the other 50% the *En* transposed to a non-linked site. What this transposing *En* illustrated was that the transposon would remain on the chromosome indefinitely, at least half of the time. And in remaining on this chromosome, *En* transposed within a linkable distance of the original site (*a1*). Since it is known that this excision leaves alterations at least most of the time, one would not be surprised to find that the footprints left by the transposon are distributed throughout chromosome 3. This *En* was called *En-vagabond.*[25]

2 GENE ALTERATIONS RELATED TO TRANSPOSONS

2.1 A High-yield Change

Following the *Ac*-induced excision of *Ds* from the shrunken-2 locus (encoding for the large subunit of the heterotetrameric starch synthetic enzyme adenosine diphosphoglucose pyrophosphorylase) a number of reversions were recovered.[26] Among the 23 excisions, 15 resulted in new amino acids sequences. In these amino acid additions to the protein, one was an additional tyrosine, one a serine, two had a two-amino acid addition. Of these two, one was tyrosine-tyrosine and the other was tyrosine-serine. The dry weight increase of the starch content of the seed of the tyrosine-tyrosine change was substantial (ranging from 11.4 to 17.8) (Figure 3). The percent change in the weight was approximately +14. In general, the revertant containing the additional tyrosine-serine increases the seed weight more than 11 to 18% with the starch content remaining constant. Thus, here was a clear change in seed weight induced by an altered form of the protein induced by a transposon excision leading to alterations in the nucleotide sequence.

AGP

inserts	AGP act	Inserts-Phenotype				Dry wt
		Phosphate Inhibition*				
		0	5	10	in E Coli	% diff
none	1.00	86	75		0.29	0
TYR	1.96					4.9
SER	1.64					24.9
TYR TYR	1.09	65	42			-
TYR SER	1.46	95	87		0.93	14.1
POTATO-AGP		-			0.1	-
POTATO-AGP#5		-			1.04	-
AGP=adenosine diphosphoglucose pyrophosphorylase						
*HK2 PO4 mM						

Figure 3. *Seed weight changes following reversions of a transposon allele at the sh2 locus*[26]

2.2 Gene Cluster Origin

After approximately five decades of research with the *R* gene, initiated by Stadler[27] in the 1930's and 1940's, an extensive analysis had been made of the kinds of changes that were determined by genetic analysis. Finally in the 1990's, the molecular biology has been able to uncover the basis of this gene cluster for the *r* gene that contains four functions.[28] The three functional genes of this complex are P, S1, and S2 and in addition an *Lc* factor. From sequencing the P, S1, and S2, it is quite obvious that the three functional genes are related in sequence. The S1 and S2 are in inverse order. The restriction-site analysis supports this contention and their sequence orientation. Further a *Doppia* transposon element was found associated with these genes that gives credence to the hypothesis that the complex arose from a series of events initiated by a transposon. In this scenario, the transposon caused the inverse order of the S1 and S2 alleles and the origin of the P component. Though this is a hypothesis it appears to be strong evidence that the origin of this cluster did have this transposon element origin.[28]

3   TRANSPOSONS AS FACTORS IN DIVERSITY

The case has been made for transposons in promoting diversity. These cases provided are not exhaustive; however, as further attempts are made to sequence genes, a number of these changes will be self evident.

A case has been presented for retrotransposons being prevalent in the grasses. These retrotransposons have been found in the flanking regions of genes. White et al.[29] proposed that these *copia*-like elements have a role in the evolution of gene structure and expression. They further suggest that these *copia*-like elements are ubiquitous and quite diverse in plants, and that these are an ancient component of plant genomes.

Recent studies[30,31] from the Bennetzen group at Purdue made a further and very exhaustive examination of the nested retrotransposons in the intergeneric region of the *Adh1*-F gene. They showed that it is composed almost wholly of retrotransposons and these are inserted within each other. This examination of about a 280-kilobase region contains 10 retro element families and they appear in further examination of these families. They appear in the genome in a wide range of frequencies that range from 10 to 30,000 copies per haploid genome. Since they account for more than 50% of the nuclear DNA of maize, and of course 60% of the *Adh1*-F region, it would appear that they are widely dispersed throughout the gene containing regions of the maize genome. With that in mind, are they a primary force in changing gene arrangement and changing the functionality of the expression of genes? That they appear in the genome is well illustrated. However, are they a driving force in changes in codogenic sequences?

That there is an effect has been well illustrated by the illustration that there is alternative splicing induced by these retrotransposons.[32] Varogona et al.[32] alleged that the retrotransposon elements found in the three waxy alleles indicate that they are important agents of spontaneous mutation in maize. Their presence in the intron does increase the intron length by approximately 40 to 60 fold according to these authors.

Now, of course, are they a major force in the extensive diversity that is found in maize. There is some evidence that they might play a minor role. The evidence is indirect. However, it is compelling. However, some indirect evidence suggests that very little mutation that targeted a wide variety of genes arises from the retroelements.

In isolation experiments, a large number of mutants have been isolated.[33,34] Recovered were over 400 separate insertions. If the retrotransposons are a factor in mutation induction, it is surprising that all of the insertions that were found were transposons and none appeared to have retrotransposons. Though this evidence is indirect it is suggested that retrotransposons are not a major factor in mutant induction and diversity in maize.

References

1.   M.M. Goodman and C.W. Stuber, *Maydica*, 1983, **28**, 169.
2.   C.W. Stuber and M.M. Goodman, 'Agriculture Research Results', U. S. Department of Agriculture, 1983, 29 pp.
3.   T. Helenjaris, D. Weber, and S. Wright, *Genetics*, 1988, **118**, 353.
4.   P.A. Peterson, *CRC Crit. Rev. Plant Sci.*, 1987, **6**, 105.
5.   H. de Vries, 'Die mutations theorie', Leipzig, Verlag Weit and Co., 1901, 63 pp.
6.   C. Correns, *Sitzb. Preus. Akad. Wiss.*, 1919, **34**, 585.
7.   R.A. Emerson, *Am. Nat.*, 1914, **48**, 87.
8.   B. McClintock, *Carnegie Institution of Washington Year Book*, 1944, **43**, 127.
9.   B. McClintock, *Genetics*, 1944, **29**, 478.
10A. B. McClintock, *Cold Spring Harbor Symp. Quant. Biol.*, 1951, **16**, 13.
10B. P.A. Peterson, *Maydica*, 1995, **40**, 117.
11.   B. McClintock, *Science*, 1984, **226**, 792.
12.   C.W. Stuber, M.M. Goodman, and F.M. Johnson, *Biochem. Genet.*, 1977, **15**, 383.
13.   M.M. Goodman, C.W. Stuber, K. Newton, and H.H. Weissinger, *Genetics*, 1980, **96**, 697.
14.   P. Franken, U. Niesbach-Klosgen, U. Weydemann, L. Marechal-Drouard, H. Saedler, and U. Wienand, *EMBO J.*, 1991, **10**, 2605.
15.   K.C. Cone, S.M. Cocciolone, F.A. Burr, anad B. Burr, *Plant Cell*, 1993, **5**, 1795.
16.   V.L. Chandler, J.P. Radicella, T.P. Robbins, J. Chen, and D. Turks, *The Plant Cell*, 1989, **1**, 1175.
17.   J. Hu, B. Anderson, and S.R. Wessler, *Genetics*, 1996, **142**, 1021.
18.   E.B. Doerschug, *Theor. Appl. Genet.*, 1973, **43**, 182.
19.   V.M. Peschke, R.L. Phillips, B.G. Gengenbach, *Science*, 1987, **238**, 804.
20.   M.G. Muszynski, A. Gierl, and P.A. Peterson, *Mol. Gen. Genet.*, 1993, **237**, 105.
21.   H. Saedler and P.A. Peterson, *EMBO J.*, 1985, **4**, 585.
22.   Zs. Schwarz-Sommer, A. Gierl, H. Cuypers, P.A. Peterson, and H. Saedler, *EMBO J.*, 1985, **4**, 591.
23.   S.R. Wessler, *Science*, 1988, **242**, 399.
24.   H. Sommer, U. Bonas, and H. Saedler, *Mol. Gen. Genet.*, 1988, **211**, 49.
25.   B.-S. Seo and P.A. Peterson, *Theor. Appl. Genet.*, 1996, **93**, 151.
26.   M.J. Giroux, J. Shaw, G. Barry, B. Greg Cobb, T. Greene, T. Okita, and L.C. Hannah, *Proc. Natl. Acad. Sci. USA*, 1996, **93**, 5824.
27.   L.J. Stadler, *Cold Spring Harbor Symp. Quant. Biol.*, 1951, **16**, 49.
28.   E.L. Walker, T.P. Robbins, T.E. Bureau, J. Kermicle, and S.L. Dellaporta, *EMBO J.*, 1995, **14**, 2350.
29.   S.W. White, L.F. Habera, and S.R. Wessler, *Proc. Natl. Acad. Sci. USA*, 1994, **91**, 11792.
30.   J.L. Bennetzen, *Trends in Microbiology*, 1996, **4**, 347,
31.   P. SanMiguel, A. Thikhonov, Y.-K. Jin, N. Motchoulskaia, D. Zakharov, A. Melake-Berhan, P.S. Springer, K.J. Edwards, M. Lee, Z. Avramova, and J.L. Bennetzen, *Science*, 1996, **274**, 765.
32.   M.J. Varogona, M. Purugganan, and S.R. Wessler, *The Plant Cell*, 1992, **4**, 811.
33.   P.A. Peterson, 'Maize Breeding and Genetics', John Wiley & Sons, NY, 1978, p. 601.
34.   P.A. Peterson, *Proc. XVI Conference of EUCARPIA*, 1993, p. 12.

Cytogenetic Investigations in *Zea mays* and *Sorghum bicolor* with Special Reference to Polyteny, B-Chromosomes and Meiosis

L. Gavrilã,[1] L. Ghetea,[1] A. Jacotã,[2] V. Scorpan,[2] M. Stefan,[1]
V.-R. Gavrilã,[1] L.-B. Gavrilã,[1] R. Basalic,[1] M. Cealâcu,[1] C. Grecu,[1]
G. Alexandru[1] and A. Petrescu[1]

[1] GENERAL GENETICS AND EVOLUTION DEPARTMENT, BUCHAREST UNIVERSITY, ALEEA PORTOCALILOR 1–3 77206, ROMANIA
[2] GENETICS INSTITUTE, ACADEMY OF SCIENCES, CHISINÃU, MOLDAVIA REPUBLIC

1. INTRODUCTION

Zea mays is an outstanding cytological and genetical instrument which allowed real advances in understanding intragenic recombination, the initiation of synapsis, functioning of the nucleolus organizer, genetic activity of the B-chromosomes and gross duplications within the corn genomes and last but not least, the presence and behaviour of mobile genetic elements (i.e. transposons) with major impacts on deciphering of eukaryotic genome organization and function [1-17, 18-23, 24].

1.1 Chromosomal complement in *Zea mays* : knobs, A- chromosomes and B-chromosomes

A peculiar cytological feature of maize chromosomal complement, taken as a marker in cytogenetical and genetical investigations in maize, is the presence of heterochromatic knobs.

Despite the presence of knobs, B- chromosomes and transposable elements, the maize has a remarkably stable karyotype.

It was Kuwada [25] to describe, for the first time, accessory chromosomes in maize, in addition to the complement of 10 pairs of chromosomes found in all studied corn plants.

Randolph [26,27] baptized these supernumerary chromosomes, B- chromosomes, in order not to be confused with usual, constant chromosomes, which are called A- chromosomes. Randolph [28] recorded 10, 40 and even 80 chromosomes for *Zea mays*, while Beadle [29] recorded 30 chromosomes in diploid chromosomal complement of Indian Corn. Maize chromosomes prepared by an enzymatic maceration method and treated with trypsin or SDS showed clear G- bands spreading along the chromosomes. While the G- banding pattern in maize is somewhat different of common G- banding pattern in animals, C-banding pattern is obviously similar. Both, G- banding and C- banding patterns allow one to identify accurately the individual chromosomes of maize [30].

The precise identification of maize individual chromosome is a necessity for gene mapping and identification of transposon insertion [31].

Fewer records on *Sorghum bicolor* chromosomal complement are available, and Darlington and Wylie's (1961) book on "Chromosome Atlas of Flowering Plants" [32] does not include the number of chromosomes of diploid complement for this species. However, in *Sorghum purpureo-sericeum* there were identified 10+1-6 B-chromosomes [32], while in *S. verticilliflorum* there is a 2n = 20+1B [32].

A regional differentiation of maize populations concerning the knobs number, has been encountered in material analysed from Romania [33]. The highest number of knobs was encountered in maize populations from West part of Romania. Giura [34] presented a survey on the B- chromosome frequency in the local populations of Romanian maize. Most of the plants exhibit 1-2 B-chromosomes, but occasionally, plants with as many as 5 B- chromosomes were recorded in the local population called "Moldovenesc de Lespezi".

1.2 Endocycles: Endopolyploidy and Politeny

Endocycles in maize have been reported only in root metaxylem where the maximum degree of endopolyploidy is 32 C [35-37]. Nagl [38] underlined that less clear polytenic structures have been observed in nuclei of the endosperm proper, in *Zea mays* [39,40], as well as in *Phaseolus coccineus* and *Ph. vulgaris* [41-46]

The level of endopolyploidy in root cap cells of maize is 16 C [47] and under infections with *Ustilago maydis* the level of endopolyploidy reached by leaf sheets is 64n [38].

Till now, polytene chromosomes in plants were described in antipodal cells, synergids, embryo suspensor, endosperm haustoria and anther hairs [38].

In the root tip cells, the region of endocycles clearly coincides with the region of cell elongation and differentiation [47-50].

The proto- and metaxylem cells change from mitosis to endomitosis earlier than the pith and cortex parenchyma cells, and show increased ploidy before their transformation into tracheary elements. A region of endopolyploidisation several times longer than in the root tip, also occurs in the shoot apex.

In *Zea mays* we describe here, for the first time, the presence of polytene chromosomes in certain types of root cell, presumably belonging to metaxylem.

1.3. Nucleolus organizer region (NOR) in *Zea* and *Sorghum*

In *Zea mays*, Mc Clintock [7] found that nucleoli arise only on chromosome 6 in a region situated on short arm which contains a large heterochromatic knob and, just distal to it, the secondary constriction. The NOR situated here is repetitive. It was estimated to be thousands of copies of the ribosomal genes localized in NORs.

1.4. Meiosis in maize

It is an usual one and was described by Rhoades [51,52]. The corn synaptonemal complex (SC) is typical in structure and time appearance [53, 54].

The investigations of Maguire et al. [55] evidenced that homozygotes for the *dsy¹* desynaptic mutant of maize show massive failure of chiasma maintenance during diplotene and diakinesis. A wider than normal SC in central region and uniform apparent weakness of central region cross connections to spreading procedures, strongly suggest the presence of a genetic lesion in a synaptonemal complex central region component. The *dsy¹* mutant may provide an especially important source of material for molecular studies on the nature of chiasma maintenance mechanism.

According to our knowledge, no studies on meiosis in *Sorghum* were performed till now, thus we present here for the first time, such studies in *Sorghum bicolor*.

2. MATERIALS AND METHODS

As biological material, two varieties of pop - corn maize (*Zea mays var. everta*), have been used, one with red kernels, and other with white kernels, as well as a hybrid labeled X_{001} and a local variety from Bertea (Prahova district). To study the *Sorghum* from a cytogenetical point of view, we have used local populations from Romania and Moldavia Republic, as well as 5 varieties from Fundulea Institute for Cereals and Technical Plants.

The kernels were kept in tap water for 24 hours, at room temperature, and subsequently were transferred to Petri dishes on wet filtered paper for germination. Prefixation, fixation, hydrolysis and staining of root meristems were performed according to Feulgen method. The microscope visualization, examination and photography were performed using a Fluoval light microscope.

For the study of meiosis, stamens of *Zea mays* and *Sorghum bicolor* were fixed in a 3:1 mixture of ethanolum absolutum and glacial aceticum acid for 72 hours and subsequently kept in refrigerator in 70^0 ethanolum. Aceto-carmine method for the staining of meiotic chromosomes was employed. The synaptonemal complexes have been evidenced using the Counce and Meyer (1973) method and Moses and Dresser method (1971, 1980). The staining of material was performed in a 50% argentum nitrate solution, 3 hours at 60^0 C.

3. RESULTS AND DISCUSSIONS

Our cytogenetic investigations allow us to evidence karyological variability of the studied forms of *Zea mays* on the background of a constant diploid complement ($2n = 20$) (Figure 1a,b).

Even mitotic chromosomes in prometaphase exhibit knobs (Figure 2). The most obvious aspect of chromosomal phenotype is concerning the chromosomes of the 6th pair which exhibit satellites (Figure 1a,b). The first pair includes the largest submetacentric chromosomes, while the pairs 2 to 5 comprise submetacentric chromosomes of medium size. The pairs 6-10 comprise subtelocentric chromosomes (Figure 1b).

The presence of B - chromosomes in maize genome is a peculiar feature of this species. In *Zea mays var. everta*, in some samples were not recognized B - chromosomes, while in others they are variable in number, from one to as much as seven (Figure 1a,b). In the variety with white kernels, the number of B - chromosomes is less than in the variety with red kernels. The B - chromosomes are smaller than A - chromosomes, usually telocentric, only occasionally meta- or submetacentric. They are more condensed compared with A-chromosomes: even in interphase, the heterochromatic blocks corresponding to B-chromosomes are obvious and distinctive from the heterochromatic sectors of A - chromosomes.

In local population of maize, the presence of B - chromosomes is a constant characteristic, while in hybrid maize , the absence of such chromosomes is typical.

The presence of satellites could be also evidenced in Metaphase chromosomes, such chromosomal markers being evidently separated from the rest of the chromosome. In some instances, one can notice the existence of acentric fragments (Figure 5), which might represent the consequence of translocations and dicentric chromosome presence. A characteristic feature of the variety IP 87 is the presence of ring chromosomes (Figure 6). We can not offer any explanation for the presence of such restructured chromosomes, except for a speculative one, involving the instability of the genome due to the presence of B - chromosomes, or of transposable elements.

There is an obvious tendency for some chromosomes to associate their satellitic regions (Figure 4).

In another variety IP_3 , heterochromatic blocks of different sizes could be visualized in an interphase nucleus and we interpret such structures as representing some chromosomal knobs of A - chromosomes, while others could represent B - chromosomes which retain their condensation during interphase. The presence of B - chromosomes in a large number in this variety is positively correlated with the appearance of mitotic abnormalities, which resemble clastogenic effects of some mutagens, such abnormalities being represented by anaphase bridges of different types (Figure 7).

These aspects of a dynamic genome of maize tell us about versatility of the genetic apparatus of *Zea mays* on the background of a constant A-chromosomal complement, explaining the impressive variability of the species and its world - scale geographical distribution.

The study of synaptonemal complexes in *Zea mays* (Figure 8) allowed us to evidence the CS profiles of the bivalents. The complement of synaptonemal complexes ha   a normal appearance. In figure 8 the bar represents $0,1\mu m$, while in upper-right triangle the bar represents $0,3$ μm.

The investigations performed in *Sorghum bicolor* took into consideration 5 varieties of this crop. The chromosomal complement of these varieties was $2n = 20$ and the presence of 1 or 2 B - chromosomes was a constant characteristic.

In *Sorghum bicolor* we have performed an extensive investigation concerning the dynamics of the chromosomes in meiosis. In Prophase I the most interesting aspect is related to the nucleolar activity. In Zygotene, the nucleolus is built at the nucleolus - organizing region of a specific chromosome;   in Pachytene (Figure 9), the nucleolus reaches its largest size. In Diplotene, one can notice the participation of more than one chromosome in nucleolar activity, while in diakinesis,   the nucleolus activity is somewhat reduced, the meiocyte preparing for disorganizing   the nucleolus itself and disrupting nuclear envelope.

Very interesting aspects were encountered in first Metaphase (M I) concerning the behaviour of B - chromosomes (Figure 10). Such chromosomes usually remain out of Metaphase I arrangement of the bivalents at the equatorium of the spindle of a dividing cell. This peculiar behaviour of B - chromosomes in meiosis will lead to the appearance of disturbances in the distribution of these B - chromosomes in the daughter cells, most of them remaining outside the daughter nuclei, even that they migrated near such nuclei, while others remained as retarded chromosomal structures at the equator   of dividing cell (Figure 11).

It is worthy to mention that nucleolar activity reappears in the second meiotic Prophase (P II) (Figure 12), while the lack of incorporation of B - chromosomes in microspore nuclei of a tetrade is a sign of random distribution of such supernumerary chromosomes from parents to offspring (Figure 13).

In conclusion, our cytogenetical investigations in *Zea mays* and *Sorghum bicolor* evidenced the architectural complexity of genome in these very important crops. The most characteristic feature of this complexity is the presence and behaviour of B - chromosomes, varying in number from 0 to as many as 7.

The meiotic behaviour of B - chromosomes explains their variability in number among different varieties, and even within a certain variety.

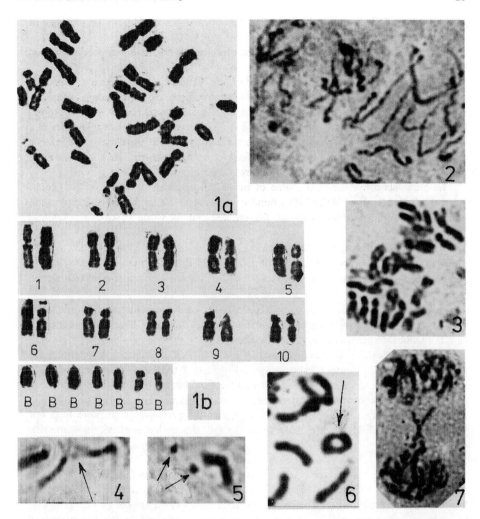

Figures 1 - 7. *Mitotic chromosomes of Zea mays (see the explanations in text)*

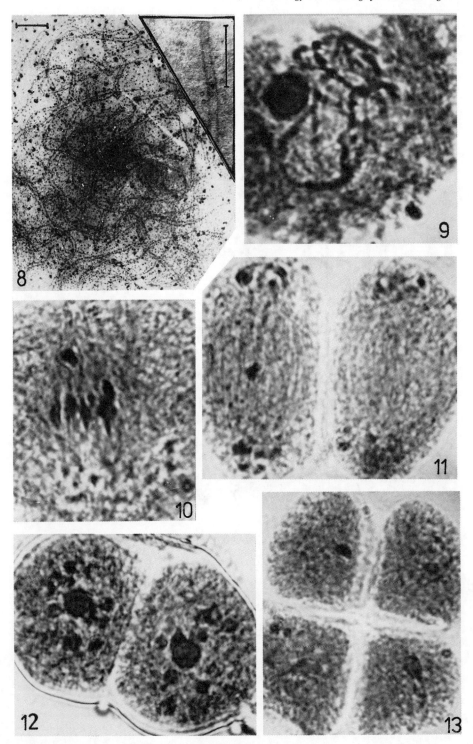

Figures 8 - 13. *Synaptonemal complex in* <u>*Zea mays*</u> *(8) and different stages of meiosis in* <u>*Sorghum bicolor*</u> *(9-13) (see explanations in text)*

In *Sorghum bicolor*, the most interesting functional aspect of meiosis is concerning nucleolar activity which reaches the highest levels in Prophase I and to a lesser extent in Prophase II. This situation is comparable with ribosomal gene amplification phenomenon which is encountered in ovogenesis, leading to the appearance of thousands of extrachromosomal nucleoli and having as output the preparing of ribosomal reserves of the next generation.

The presence of B-chromosomes both in *Zea mays* and *Sorghum bicolor* could be exploited by scientists and breeders to build artificial chromosomal structures used as vectors in the process of directed gene transfer in cultivated plants in order to obtain transgenic plants. From this point of view, it is decisive to know the behaviour of such entities, in order to ensure their isolation and manipulation in experimental work.

The presence of B-chromosomes in the way between the poles of a dividing cell, as we proded to be the case in *Sorghum bicolor* Ana-Telophase I, will facilitate their isolation.

Once being isolated, a B-chromosome can be used in amplification procedure in artificial reconstruction of chromosomal structures, used as vehicles in gene tranfer processes.

3.1 Polytene chromosomes in *Zea mays*

We have obtained conclusive data on the existence of polyteny in certain cell types of root tip in *Zea mays* (Figures 14-20). A presumptive sequence of events leading to polyteny is as follows: the cells that have to develop potyteny, enter firstly endocycle, leading to endopolyploidy. The nuclei of these cells enlarge, becoming 10-20 times larger than mitotic diploid nuclei (which develop mitotic cycle) (Figure 14). Such endopolyploid nuclei (Figures 15,17,18), reaching a certain level of endopolyploidy, stop the endomitotic cycle and switch to a polyteny endoreduplication. Occasionally, the homologous chromosomes undergo a type of somatic pairing and this event could be readily accomplished since these homologous chromosomes, as a consequence of endopolyploidy, are numerous and in intimate vicinity, as we have found in mitotic somatic cells (see Figure 3). Somatic pairing of such endochromosomes is a prerequisite for generating typical aspects of polyteny, with a characteristic banding pattern. The polytene chromosomes take specific aspects of cable-like threads (Figures 15-20) composed of multiple endochromosomes, reflecting the structure of interphase chromosomes on a large scale[38]. The degree of bundling of the endochromosomes is in direct connection to the level of endopolyploidy reached by the nucleus.

The euchromatic sectors of these polytene chromosomes have a mainly granular structure because the corresponding chromosomes of the endochromosomes do not match up to form bands. But, occasionally, a few bands can be sectorially organized, being in no respect different from salivary glands or embryo suspensor polytene chromosome bands[56].

The heterochromatic portions appear either in the form of compact and sometines vacuolated blocks or in the form of groups or large granules, generating heterochromomeres[38].

Zea mays root polytene chromosomes range from an obviously bundled state to one in which the euchromatic segments of the endochromosomes radiate star-like from the endochromocentres (Figure 16). Instances in which in one and the same polyploid nucleus, some endochromosomes exhibit polytenic phenotype while others remain as interphase chromatin threads, are frequent.

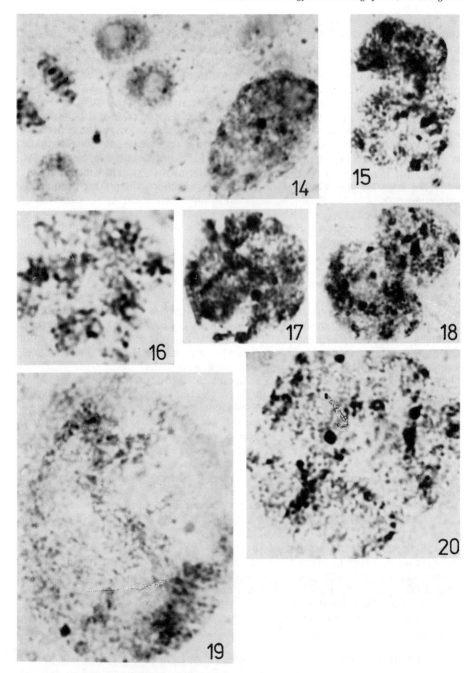

Figures 14 - 20. *Polyteny in* Zea mays *root tip cells. Note the difference between mitotic nuclei (left side) and an endopolyploid nucleus (down right side) in figure 14. See the text for further explanations*

The functional significance of the polyteny in root tip cells of *Zea mays* can be connected to a necessity for a high level of nucleolar material (i.e. ribosomal precursor) production. This is a kind of cell functioning specialization in a way in which root meristematic cells are engaged in mitotic divisions with a high speed of nucleus and cytoplasm division, while same root tip cells differentiating the endopolyploid / polytenic cycle, are engaged in high rates of ribosomal precursor synthesis. The polytenic cells help the organism in two directions: 1- assuring the lengthening of the root by the enlargement, such cells reaching large sizes; 2- functioning as nurse-like cells for mitotic ones, their ribosomal precursors are transferred in some way into mitotic cells, these later ones being engaged essentially in division and less in cell syntheses. This speculative explanation is also based on the necessity for this crop to develop rapidly a powerful root system, since maize is cultivated early in the spring season in which weather changes are frequently encountered.

In *Zea mays*, the root polytene nuclei are undoubtedly involved in high levels of ribosomal precursors synthesis (i.e. nucleolar building). All these polytene nuclei exhibit large Feulgen regative areas, representing the nucleoli locations (Figures 19,20). Some 4 to 10 such areas are easily recognized. In *Zea mays*, the most obvious polytene chromosome is the nucleolus organizing chromosome (Figures 15,17,18,19). More than one nucleolus organizing chromosome participates in the building of the same nucleolus, a situation which Gavrilâ and Mihâescu[57] have encountered also in root nodule polytene nuclei in *Trifolium pratense* .

The final stage of polytene nuclei in *Zea mays* reveals a kind of emptied organelle, ribosomal precursors being released into the cytoplasm and from here transferred to other cells. In such nuclei, trajectories of polytene chromosomes are more clear, but their receptiveness for the dyes became very weak. Such nuclei are prepared for death (Figure 19).

This is the very fate of all cells which develop polyteny. They are a pure image of an ephemeral and altruistic file style.

References

1. B.McClintock, "Chromosome morphology in *Zea mays* ", *Science*, 1929a, **69**, 629

2. B.McClintock, "A cytogenetical and genetical study of triploid maize", *Genetics*, 1929b, **14**,180-222

3. B.McClintock, "Cytogenetical observations of deficiencies involving known genes, translocations and an inversion in *Zea mays*", *Missouri Agric.Exp.Stn. Res.Bull.*, 1931, **163**,1-30.

4. B.McClintock, and H.E.Hill, "The cytogenetical identification of the chromosome associated with the R-G linkage group in *Zea mays* ", *Genetics*, 1931, **16**,175-190.

5. B.McClintock, "A corellation of ring-shaped chromosomes with variegation in *Zea mays* ", *Proc.Natl.Acad.Sci.*, USA, 1932, **18**, 677-681

6. B.McClintock. "The association of nonhomologous parts of chromosomes in the mid-prophase of meiosis in *Zea mays* ", *Z.Zellforsch. Microskop. Anat.*, 1933, **19**,191-237.

7. B.McClintock, "The relation of a peculiar chromosomal element to the development of a nucleoli in *Zea mays*", *Z.Zellforsch. Microskop. Anat.*, 1934, **21**, 294-328.

8. B.McClintock, "The fusion of broken ends of sister half-chromatids following chromatin breakage at meiotic anaphases", *Missouri Agric.Exp.Stn.Res.Bull.*, 1938a, **290**,1-48.

9. B.McClintock, "The production of homozygous defficient tissues with mutant characteristics by means of aberant mitotic behaviour of ring-shaped chromosomes", *Genetics*, 1938b, **23**,315-376.

10. B.McClintock, "The behaviour in successive nuclear division of a chromosome broken at meiosis", *Proc.Natl.Acad.Sci.*, USA, 1939, **26**,405-416.

11. B.McClintock, "The stability of broken ends of chromosomes in *Zea mays*", *Genetics*, 1941, **26**,234-282.

12. B.McClintock, "The origin and behaviour of mutable loci in maize", *Proc.Natl. Acad.Sci.* USA,1950, **36**,344-355.

13. B.McClintock, "Chromosome organization and genic expression", Cold Spring Harbor Symp.Qant.Biol., 1951, **16**,13-47.

14. B.McClintock, "Controlling elements and the gene", Cold Spring Harbor Symp.Qant.Biol., 1956, **21**,197-216.

15. B.McClintock, "Chromosome constitutions of Mexican and Guatemalan race of maize", Carnegie Inst. Wash. Yearb., 1959, **59**,461-472.

16. B.McClintock, The control of gene action in maize", Brookhaven Symp. Biol., 1965, **18**, 162-184.

17. B.Mc Clintock, "Genetic systems regulating gene expression during developement", Devel.Biol.Suppl., 1968, **1**,84-112.

18. W.R.Carlson, "A test of homology between the B-chromosome of maize and abnormal chromosome 10, involving the control of nondisjunction in B's", *Mol.Gen.Genet.*, 1969, **104**,59-65.

19. W.R.Carlson, "Nondisjunction and isochromosome formation in the B chromosome of maize", *Chromosoma*, 1970, **30**,356-365.

20. W.R.Carlson, "Instability of the maize B chromosome", *Theor.Appl.Genet.*, 1973a, **43**,147-150.

21. W.R.Carlson, "A procedure for localizing genetic factors controlling mitotic nondisjunction in the B chromosome of maize", *Chromosoma*, 1973b, **42**,127-136.

22. W.R.Carlson, "B chromosomes induce nondisjunction in 9B[9] pollen", *Maize Genet.Coop. News Letter*, 1974a, **42**,81-83.

23. W.R.Carlson, "A mutant of cromosomal behaviour in mitosis and meiosis", *Maize Genet.Coop. News Letter*, 1974b, **48**,76-80.

24. W.R.Carlson, "The Cytogenetics of Corn", Corn and Corn improvement, Academic Press, New York, 1974, 225-304.

25. Y.Kuwada, "Meiosis in the mother cells of *Zea mays* ", *Bot.Mag.* (Tokyo), 1911, **25**,163-181.

26. L.F.Randolph, "Chromosome numbers in *Zea mays* L., Cornell Univ. Agric. Exp.Stn.Mem., 1928a, 117.

27. L.F.Randolph, "Types of supernumerary chromosomes in maize", *Anat.Rec.*, 1928b, **41**,102.

28. L.F.Randolph, "Some effects of high temperature on polyploidy and other variations in maize", *Proc.Natl.Acad.Sci.*, USA, 1932, **18**,222-229.

29. G.W.Beadle, "Genetical and Cytogenetical Studies of Mendelian Asynapsis in Zea mays", Cornell Univ.Agric.Exp.Stn.Mem., 1930, **129**,1-23.

30. K.Kakeda, H.Yamagata, K.Fukui, Z.A.Wei and F.S.Zhu, "High resolution bands in maize chromosomes by G-banding methods", *Theor.Appl.Genet.*, 1990, **80**,265-272.

31. S.Dash, P.A.Peterson, "Chromosome constructs for transposon tagging of desirable genes in different parts of Zea mays genome", *Maydica*, 1989, **34**,247-261.

32. C.D.Darlington, A.P.Wylie,"Chromosome Atlas of Flowering Plant", George Allen & Unw in Ltd, London, 1961.

33. M.Cristea, "Rasele de porumb din Ronania", Ed.Acad.R.S.R., Bucuresti, 1977.

34. A.Giura, "The cytological study of some heterochromatic elements of maize genome. I. The B-chromosomes in Romanian local forms of maize", Probl. Genet.Teoret.Aplic., 1988, XX, **3**,151-159.

35. H.Swift, "The constancy of DNA in plant nuclei", *Proc.Natl.Acad.Sci.*, USA, 1950, **36**,643-645.

36. A.List, "Some observations on DNA content and nuclear volume growth in the developing-xylem cells of certain higher plants", *Am.J.Bot.*, 1963, **50**,320-329.

37. V.Lai, L.M.Srivastaba, "Nuclear changes during differentiation of xylem vessel elements", *Cytobiologie*, 1976, 12,220-243.

38.   W.Nagl, "Endopolyploidy and Polyteny in Differentiation and Evolution, Towards an Understanding of Quantitative and Qualitative Variation in Nuclear DNA in Ontogeny and Phylogeny", North-Holland Publishing Comp. Amsterdam, New York, Oxford, 1978.

39.   E.Tschermak-Woess,   U.Enzenberg-Kunz,   "Die   Struktur   der hochendopolyploiden kerne im Endosperm von *Zea mays*, das auffhallende Verhalten ihrer Nukleolen und ihr Endopolyploidiegrad", *Planta*, 1965, **64**,149-169.

40.   J.Stephen, "Occurrence of polyteny, endopolyploidy and numerical vaiation of nucleoli in maize endosperm", *Science and Culture* (India), 1973, **39**,323-324.

41.   W.Nagl, "karyologische Anatomie des Endosperms von Phaseolous coccineus, *Osterr.Bot.Z.*, 1970, **118**,566-571.

42.   W.Nagl, "Spontane Fragmentation Endopolyploider Kerne in Dauergeweben von Allium-Arten, *Oster.Bot.Z.*, 1970a, **118**,431-442.

43.   W.Nagl, "Differentielle RNS-Synthese in pflanzlichen Riesenchromosomen", *Ber.Dtsch.Bot.Ges.*, 1970b, **83**,301-309.

44.   W.Nagl, "Inhibition of polytene chromosome formation in Phaseolus by polyploid mitosis", *Cytologia*, 1970c, **35**,252-258.

45.   W.Nagl, "Karyologische Anatomie des Endosperms von Phaseolus coccineus", Osterr.Bot.Z., 1970d, **118**,566-571.

46.   W.Nagl, "Gibberellinsaure-stimulierte Genaktivitat im Endosperm von Phaseolus", *Planta*, 1971, **96**,145-151.

47.   P.W.Barlow, "The root cap In: The Development and Function of Roots (J.G.Torrey and D.T.Clarkson eds.)", Academic Press, New York and London, 1975, 21-54.

48.   F.D'Amato, "Polyploidy in the differentiation and function of tissue and cells in plants: a critical examination of the literature", *Caryologia*, 1952a, **4**,311-358.

49.   F.D'Amato, "New evidence on endopolyploidy in differentiated plant tissues", Caryologia, 1952b, **4**,121-144.

50.   W.Nagl, "Der mitotische und endomitotische Kernyklus bei Allium carinatum. I. Struktur, Volumen und DNS-Gehalt der Kerne", *Osterr.Bot.Z.*, 1968, **115**,322-353.

51.   M.M.Rhoades, "Meiosis in maize", *J.Hered.*, 1950, **41**,58-67.

52.   M.M.Rhoades, "The cytogenetics of maize (in:) Spraque G.F. (ed) Corn and Corn improvement", Academic Press, New York, 1955, 123-219.

53. A.G.Underbrink, Y.C.Ting, A.H.Sparrow, "Note on occurrence of a synaptinemal complex at meiotic prophase in *Zea mays*", *Can.J.Genet.Cytol.*, 1967, **9**,606-609.

54. C.B.Gilles, "Ultrastructural analysis of maize pachytene karyotypes by three dimensional construction of the synaptonemal complexes", *Chromosoma*, 1973, **43**,145-176.

55. M.P.Maguire, R.W.Riess, A.M.Paredes, "Evidence from a maize desynaptic mutant points to a probabole roe of synaptonemal complex central region components in provision for subsequent chiasma maintenance", *Genome*, 1993, **36**,797-807.

56. W.N agl, "Banded polytene chromosomes in the legume Phaseolus vulgaris", *Nature* (London), 1969, **221**,70-71.

57. L.Gavrila, Gr.Mihaescu, "Nucleolus organizing chromosomes and ribosomal gene amplification phenomenon in root nodule cell of Trifolium pratense: an electronomicroscopic study", *Cytologia* (Tokyo), 1992, **57**,15-20.

Structural and Functional Organization of Maize Chromatin During Embryo Germination

E. I. Georgieva

DEPARTMENT OF MOLECULAR GENETICS, INSTITUTE OF GENETICS, BULGARIAN
ACADEMY OF SCIENCES, 1113 SOFIA, BULGARIA

1 INTRODUCTION

The huge amount of eukaryotic DNA as well as the other nuclear components must be organized in a tiny nucleus in a highly ordered way to achieve a perfect synchronization and accuracy of the various structural/functional processes occurring in chromatin. The basic structural chromatin subunit is the nucleosome which consists of about 160 bp of DNA wrapped in approx. two turns around two molecules of each of the four core histones (H2A, H2B, H3 and H4). In both animals and plants, DNA is arranged in a 10 nm nucleosomal fiber which is further wound in a 30 nm solenoid[1]. At the next organizational level, the DNA is attached to the nuclear matrix[2]. Despite the tremendous progress that has been made in defining the steps of chromatin and nuclear assembly, our knowledge is still far from an exact understanding of how this complex chromatin structure can, under tight control, modulate the regulatory mechanisms through development of organisms. Nowadays one of the most challenging problems of molecular biology is the understanding of how genes are activated from a repressed into an actively transcribed state.

During the last few years we studied the structural and functional relationships in plant genome using germinating plant embryos as a model system. The germination of plant embryo is a differentiation process in which the cell nucleus undergoes changes from an extremely low metabolism in quiescent embryos to a state of fully restored activity after the start of seed imbibition. This process therefore represents a sequence of molecular events that transform the heterotrophic embryo into a complex autotrophic organism. Hence, germinating plant embryos are a good, natural system for unraveling the mechanisms maintaining and regulating structural and functional organization of plant genome.

The present report summarizes our investigations on the mechanisms involved in the regulation and organization of maize chromatin during the embryo germination process.

1.1 Cell Cycle Analysis

Studies of cellular and molecular events during plant embryo germination require a detailed analysis of nuclear processes and cell cycle parameters. In view of the limited and often controversial data we set out to study the cell cycle distribution of maize embryo cells at different time points of germination process by DNA flow cytometry[3,4] and some of the factors involved in the molecular control of the cell cycle as: thymidine kinase activity, in vivo transcriptional activity of chromatin, nuclear proto-oncogene and p53 tumor suppressor-related proteins[5], fluctuation of histone acetyltrasferase and histone deacetylase enzyme activities[6].

Figure 1 shows the size of the maize seedlings from 0 to 72 hours after the start of imbibition used for the analysis of the cell cycle. Nuclei were isolated at 12 time points in the germination period and stained with propidium iodide (central part of the figure). For the planned study it has been used seeds from Zea mays, strain M320, created at the Institute of Genetics in Sofia. Maize embryos from dry seeds at 0 to 24-30 hours were collected manually to avoid the presence of endosperm cells.

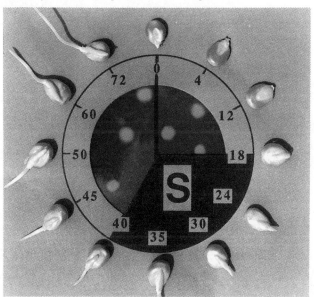

Figure 1 *Schematic presentation of maize germination process and the size of the seedlings used for nuclei isolation. Embryos and seedlings of the same size were always selected in order to minimize deviation due to different levels of cell differentiation.*

1.1.1 DNA Flow Cytometry. Purified maize nuclei were stained with propidium iodide and run on a FACScan-immunoflowcytometer. The analyzes of DNA distribution by flow cytometry[3,4] showed that the transition from early to late stages of germination was accompanied by a decrease of G_1 nuclei and a dramatic increase of G_2 nuclei at 72 hours.

The results of flow cytometric analyses of 12 time points during the germination of maize embryos with respect to the percentage of cells in S-phase, G_1 and G_2 are summarized in Table 1. At the start of germination (dry embryo) the percentage of cells in S-phase amounted to approx. 30% and this value decreased rapidly within the initial 4 h of germination to about 15%. From 12 to 72 h of germination there was an increase in the percentage of cells in S-phase. The increase was not continuous but showed one not well pronounced peak until 30-40 h, then a plateau until 60 h and a second pronounced increase between 60 and 72 hours. These data show a biphasic distribution of S-phase, one peak around 24-35 h and a second peak after 60 h of embryo germination. The G_1 population increased substantially within the initial 4 h of the germination period from 60% at the start of imbibition to 75% at 4 h and remained almost constant up to 12 h. Between 30 and 60 h there was only a minor decrease in the G_1 population. The transition from 60 to 72 h was again accompanied by a rapid decrease of the G_1 cells to reach a percentage of less than 25%. The population of G_2 nuclei showed an almost constant level until 18 h after the start of imbibition and between 30 and 60 h. One increase in the cells in G_2 population was detected between 24 and 30 h of embryo germination and a second one, between 60 and 72 h.

1.1.2 Thymidine Kinase Activity.   To characterize the validity of this results and the onset of S-phase we measured the activity of thymidine kinase, an enzyme tightly associated with S-phase[4,7] in the same embryo material that was used for isolation of nuclei. It is well established that during cell proliferation thymidine kinase is induced concomitant with the onset of DNA synthesis and its expression stopped at the end of DNA replication.

The activity of thymidine kinase was at a low but measurable level at the start of germination, then decreased transiently during the progression to 12 h, but increased dramatically between 12 and 30 h. Between 60 and 72 h the enzyme activity increased almost two fold. Our results shows that the sharp increase in thymidine kinase activity correlates well with the distribution of the cells in the S-phase (Table 1). Cell cycle distribution of cells and changes in thymidine kinase activity were reproducible within repeated experiments.

Table 1 *Distribution of the percentage of nuclei in S-phase, G_1 and G_2 -period and changes in the activity of thymidine kinase during the germination of maize embryos*

Time of Germination (h)	Nuclei in S-phase (%)	Nuclei in G_1 (%)	Nuclei in G_2 (%)	Thymidine Kinase Activity (pkat.g^{-1})
0 h (dry embryo)	30	60	10	0,5
4 h	15	75	10	0,5
12 h	18	72	10	0,25
18 h	22	65	13	0,5
24 h	25	55	20	1,0
30 h	28	50	20	1,1
35 h	32	50	18	1,0
40 h	35	45	20	0,9
45 h	35	45	20	0,8
50 h	36	45	18	0,8
60 h	39	42	18	1,0
72 h	44	25	25	1,5

1.1.3 In vivo Transcriptional Activity of Chromatin.   The data of in vivo transcriptional activity of chromatin are displayed in Figure 2. Analysis of total RNA synthesis by radioactively labeled uridine incorporation throughout the germination pathway revealed two peaks - at 24 and 72 h which means that an active RNA synthesis takes place at 24 and 72 h of maize embryo germination. The results obtained are consistent with the DNA flow cytometry and the thymidine kinase activity.

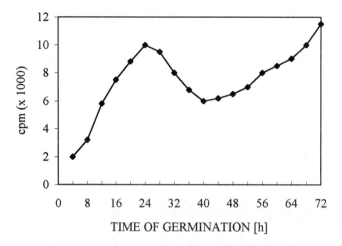

Figure 2 *Transcriptional activity of chromatin at different stages of seeds germination. The experiment was done by in vivo labeling with radioactive precursor ³H-uridine. The data are averaged from nine independent experiments*

Taking into account all of our data we can conclude: i) - within the initial 60 h of maize embryo germination cells traverse one cell cycle, ii) - the vast majority of cells approx. 75% start germination in G_1 period, iii) - only 30% of the cells enter S-phase as a partially synchronized population, iv) - the remaining cells will not enter the cell cycle during the germination period, which makes sense with respect to the diverse developmental functions of these cells. The data are also in line with our previous results[6] in which we showed that histone acetyltransferase B, an enzyme associated with the acetylation of newly synthesized histone H4 and hence with DNA replication, exhibits a biphasic fluctuation during maize embryo germination, which is in accordance with the pattern of thymidine kinase activity and with DNA flow cytometry data.

The results presented above provide the basis for functional correlation of nuclear metabolism to basic cell cycle parameters. For that reason we studied the levels of nuclear proto-oncogenes and *p53* - related proteins on RNA and protein levels during germination period[5]. Surprisingly, we found that the levels of proteins related to the *myc* family (c-MYC and N-MYC) and to the p53 homologue change in opposite ways during embryo germination. The *p53*-related protein disappears after the start of replicative DNA synthesis, whereas c-MYC and N-MYC homologues reach high levels at 24 h.

We propose that the p53-related protein can serve as a marker for the quiescent stage of the maize embryo and somehow is involved in the stabilization of the G_1/G_0 state of cells during the initial phase of germination. c-MYC and N-MYC homologues seem to be associated with the proliferative activity of maize embryo.

1.2 DNase I Sensitivity of Chromatin

Considerable progress has been made in examining the structure of DNA and the consequences of changes of its conformation by using endogenous nucleases - micrococcal nuclease, DNase I, DNase II and different restriction enzymes as probes. It is generally believed that the altered conformation of transcriptionally active sequences renders them preferentially sensitive to nucleases, particularly to DNase I. In the present research we studied general DNase I sensitivity on chromatin structure during the process of maize embryo germination. Our results present evidence for both enhanced DNase I sensitivity and dynamic changes occurring in the structure of maize chromatin core particles during transition of the fully inactive genome state (dry embryo) to the active one (24 and 72 h). From the electrophoretic analysis of nuclease-released DNA fragments, our data showed that DNase I cut the DNA in different ways at the investigated stages of embryo germination (Figure 3).

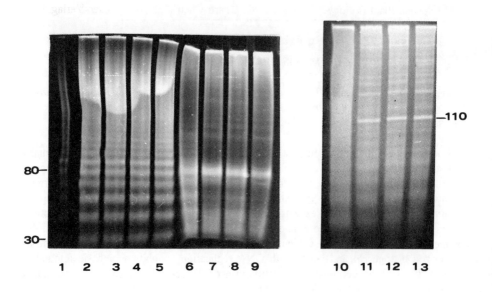

Figure 3 *Electrophoretic patterns of DNA fragments produced by DNase I digestion in denaturation conditions. Lanes 1 and 10 - 10-base nucleotide repeat used as a control; Lines 2-5 - nuclease-released DNA fragments at 5, 10, 15 and 30 min. after DNase I digestion of dry embryo chromatin; Lines 6-9 nuclease-released DNA fragments at 5, 10, 15 and 30 min. after DNase I digestion of 24 h embryo germinated chromatin; Lines 11-13 - DNA fragments after DNase I digestion of 72 h embryo germinated chromatin at 5, 10 and 30 min. DNase I treatment*

DNA fragments from dry embryo chromatin showed normal electrophoretic behavior in comparison with the control, while in the other two germinated stages (24 and 72 h) we observed differences in the pattern of DNase I accessibility. Nuclease released DNA fragments, obtained from 24 h chromatin, indicate a pause of digestion around 80th nucleotide, while in those from 72 h such a pause appears around 110th nucleotide. The observed differences in the pattern of 10-base ladder display differences in the protection of the cutting sites, which means that some sort of remodeling of maize genome takes place during transition of the cells in different stages of the cell cycle and that gross changes in nucleosomal organization accompany the onset of maize germination.

References

1. K. E. van Holde, "Chromatin", 1989, New York: Springer-Verlag.
2. S. M. Gasser, "Architecture of Eukaryotic Genes", 1988, G. Kahl, ed. (Vienna: VCH Verlagsgeselschaft), p. 461.
3. C. S. Brown, C. Bergounioux, S. Tallet and D. Marie, "A Laboratory Guide for Cellular and Molecular Plant Biology", I. Negrutiu, G, Gharti-Chetri, eds. Birkhauser-Verlag, Basel, 1991, p. 326.
4. E. I. Georgieva, G. Lopez-Rodas, A. Hittmair, H. Feichtinger, G. Brosch and P. Loidl, *Planta*, 1994, **192**, 118.
5. E. I. Georgieva, G. Lopez-Rodas and P. Loidl, *Planta*, 1994, **192**, 125.
6. E. I. Georgieva, G. Lopez-Rodas, R. Sendra, P. Grobner and P. Loidl, *J. Biol. Chem.*, **266**, 18751.
7. P. Grobner and P. Loidl, *Biochim. Biophys. Acta*, 1982, **697**, 83.

Genomic Organization and Physical Distribution of *Tyl-COPIA*-like Retrotransposons in Maize and Sorghum

A. Katsiotis,[1,2] M. Hagidimitriou[1,2] and J. S. Heslop-Harrison[2]

[1] AGRICULTURAL UNIVERSITY OF ATHENS, MOLECULAR BIOLOGY LABORATORY, IERA ODOS 75, 118 55 ATHENS, GREECE
[2] JOHN INNES CENTRE, KARYOBIOLOGY GROUP, COLNEY LANE, NORWICH NR4 7UH, UK

1. INTRODUCTION

A very large proportion of most eukaryotic genomes is composed of repetitive DNA. Studying the genomic and chromosomal organization of repetitive DNA can provide information about plant evolution and nuclear architecture; this knowledge can help to better understand relationships between species and in manipulating the flow of desirable traits from the wild species to the cultivated ones. It can also provide us with information about the chromosomal structure and the genomic organization for comparative mapping and collinearity studies between species and genera.

Reassociation kinetic studies of DNA can provide us, within approximation, with the percentage of DNA that is repetitive within a genome. Such studies revealed that repeated DNA contributes about 70% of barley,[1] 75% of rye,[1] 75% of wheat,[1] 75% of oat,[1] 52% of rice, and 68% of maize[2] total DNA content. Three major DNA kinetic classes are present in maize: a rapidly reassociated DNA that corresponds to highly repetitive DNA (present at an average reiteration frequency of 800,000 copies and contributing about 20% of the genome); a fast reassociated DNA that corresponds to mid-repetitive DNA (present at an average reiteration frequency of 1,000 copies and contributing about 40% of the genome); and a slow reassociated DNA that corresponds to the low repetitive DNA and unique sequences.[2]

Organization of repetitive DNA can be either in tandem arrays or dispersed through the genome. A major fraction of the dispersed repetitive DNA in eukaryotes has been suggested to be mobile elements or their remnants. The two types of mobile elements are the transposons (discrete sequences that have the ability to move directly from one site of the genome to another) and the retrotransposons (the DNA is transcribed into RNA, and then reverse-transcribed into DNA, which is inserted at a new site in the genome).[3] Because of the nature of their transposition, transposons are present in low copy numbers in the genomes, while retrotransposons are present in higher copy numbers.

2. *Tyl-COPIA*-LIKE RETROTRANSPOSONS IN GRAMINAEA

The *Tyl-copia*-like retrotransposons have two conserved regions at their reverse transcriptase (*rt*) and integrase (*int*) genes.[3] Primers amplifying part of the *rt* gene, corresponding to peptide sequences TAFLHG and YVDDML have been constructed, and were used in polymerase chain reactions (PCR). In most plant species investigated thus far *Ty1-copia*-like retrotransposons are ubiquitous and contibute a significant portion of plant genomes.[4,5]

For the purpose of the present experiment DNA from ten grasses belonging to different tribes was used: maize and sorghum from the Andropogonanae, paspalum from the Panicanae, wheat, barley and rye from the Triticeae, bromus from the Bromeae, oats and arrhenatherum from the Aveneae, and rice from the Oryzeae. The PCR reactions were denatured at 94°C for 1 min, followed by 35 cycles of 1 min at 94°C, 1 min at 38°C, and 1 min at 72°C, with a final elongation step of 5min at 72°C. The products were reamplified by a second PCR reaction under the same condiotions. All ten grasses gave a PCR product of the expected size, about 280 pb (Figure 1). Although the annealing temperature for the PCR cycles was low, 38 to 42 degrees, no other fractions, other than the expected, were obtained. However, each product corresponds to a pool of different *Ty1-copia*-like retrotransposon families. From previous experiments in a number of genera, cloned fragments from this single band showed heterogeneity in nucleotide and predicted peptide sequence within each genus.[5]

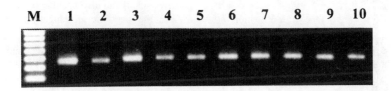

Figure 1 *PCR products obtained after the second amplification of the* rt *gene from total genomic DNA of maize (lane 1), sorghum (lane 2), paspalum (lane 3), wheat (lane 4), barley (lane 5), rye (lane 6), bromus (lane 7), arrhenatherum (lane 8), oats (lane 9), and rice (lane 10). The marker (M) is a 100 bp ladder.*

3. GENOMIC ORGANIZATION OF *Ty1-COPIA*-LIKE RETROTRANSPOSONS IN MAIZE AND SORGHUM

In order to study the genomic organization of the maize and sorghum PCR products, total genomic DNA from the ten genera used for the PCR reactions was digested with *Hae*III. Using southern hybridization as a probe the total PCR product from the maize Rt amplification was hybridized on all genera, but the strongest hybridization signal was on maize DNA (Figure 2A). The resulting smear on the maize lane, with superimposed larger fragments, suggests a dispersed organization of maize *Ty1-copia*-like retroelements in the maize genome. Hybridization signal on more distantly related genera indicates the conservation of some *Ty1-copia*-like retrotransposons throughout the grass family. Between more closely related genera like wheat, rye, and barley, common bands are present at about 0.4, 0.6, 0.9, 1.5, and 2.0 kb, indicating similar genome organization. Similar results were obtained when the sorghum Rt amplification product from PCR was hybridized to a similar filter (Figure 2B). The organization of the sorghum *Ty1-copia*-like retroelements is dispersed throughout the sorghum genome.

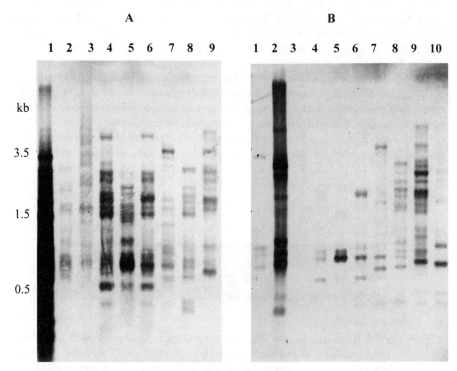

Figure 2 *Southern hybridization of maize (A) and sorghum (B)* Ty1-copia-*like retrotransposon PCR products on total genomic DNA of maize (lane 1), sorghum (lane 2), paspalum (lane 3), wheat (lane 4), barley (lane 5), rye (lane 6), bromus (lane 7), arrhenatherum (lane 8), oats (lane 9), and rice (lane 10), digested with* HaeIII *enzyme.*

Cross-hybridization of the sorghum and maize *Ty1-copia*-like retroelements with genomic DNA from other Graminaea tribes in the Southerns, are in agreement with a dendogram (Figure 3) which has been obtained from sequences derived from PCR products and described retroelements. BARE-1 and PREM-2 are two retroelements that have been studied in barley[6] and maize[7] respectively, while the rest of the sequences are cloned products from PCR.[4, 5] Because the present sequences are only a small fraction of the different *Ty1-copia*-like retroelement families present in the Granimaea family, tribes that are not closely related evolutionarily are clustered;for example wheat is clustered with oats, and not with barley; and different *Ty1-copia*-like sequences from maize are clustered with barley and rice, and not with themselves.

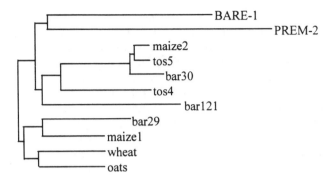

Figure 3 *Dendogram showing the relationship of* Ty1-copia-*like retrotransposons of barley (BARE-1, bar29, bar30, bar121), maize (PREM-2, maize1, maize2), rice (tos4, tos 5), wheat, and oats.*

4. PHYSICAL ORGANIZATION OF *Ty1-COPIA*-LIKE RETROTRANSPOSONS IN MAIZE

From the Southerns it is evident that retrotransposons are dispersed sequences; however, their distribution on chromosomes can be visualized by *in situ* hybridization. *In situ* hybridization of the pooled PCR RT maize probe on maize chromosomes shows that retrotransposons are present on all chromosomes with the same signal intensity without concentrating on any specific chromosome or chromosome site. Retrotransposons are absent from the centromeric and subtelomeric regions, which are important chromosome functional regions, as well as the heterochromatic and NOR regions (Figure 4).

A

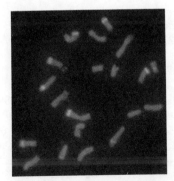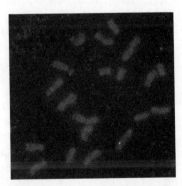

Figure 4 *Localization of* Ty1-copia-*like retrotransposon sequences on root-tip metaphase maize chromosomes by fluorescence in situ hybridization. Chromosomes stained with DAPI (A) and probed with the pooled PCR product from the maize Rt amplification (B).*

Interphase nuclei give the same results showing a uniform distribution of retrotransposons with no clustering. So retrotransposons have the ability to be inserted on all maize chromosomes and at any site of the chromosome except the previously mentioned sites. In a similar study the pooled PCR product from the *Paspalum* Rt amplification was used to probe *Paspalum* root-tip metaphase chromosomes. Again a uniform ditribution was evident, with no hybridization on centromeres and NORs. However, no differentiation between the different *Papsalum* genomes was possible, as has been previously described for the oat genomes.[8] *Ty1-copia*-like retrotransposons have been physically mapped for a number of tribes of the Gramineae family.[8, 9] Although the physical organization of *Ty1-copia*-like retroelements is not the same for all plant families and genera,[9] all grasses studied thus far show the same physical distribution of these elements on their chromosomes.

It has been shown that retrotransposons are present in numerous plant species comprising several percent of the total DNA content.[3, 4, 6, 7] Because of the high copy numbers of retrotransposon families in plants, they are considered a major contributor in generating genetic biodiversity, and can be used for phylogenetic relationships in crop species. The degree of sequence divergence of most retrotransposon families between species has been shown to be proportional to the to their evolutionary distance. Thus, studying the evolution of some vertically trasmitted retrotransposon elements can provide clues about the evolution of the Graminaea family tribes. Some retrotransposon elements that are not found outside their own genus and seem to have a recent origin, can provide speciation information within genera. Other retrotransposon elements show homology with very distant families or even taxa, indicating a horizontal transmission; however, the hypothesis of horizontal transmission is still waiting to be proved. The last few years studies have been initiated to search for the significance of retrotransposons in plant evolution and speciation, and to study their expression, regulation, and the different modes of their transmission.

References

1. R.B. Flavell, M.D. Bennett, J.B. Smith and D.B. Smith, *Biochem. Genet.*, 1974, **12**, 257.
2. S. Hake and V. Walbot, *Chromosoma (Berl.)*, 1980, **79**, 251.
3. M.A. Grandbastien, *Trends Genet.*, 1992, **8**, 103.
4. A.J. Flavell, *Genetica*, 1992, **86**, 203.
5. H. Hirochika and R. Hirochika, *Jpn. J. Genet.*, 1993, **68**, 35.
6. I. Manninen and A.H. Schulman, *Plant Mol. Biol.*, 1993, **22**, 829.
7. M.P. Turcich, A. Bokhari-Riza, D.A. Hamilton, C. He, W. Messier, C. Steward and J.P. Mascarenhas, *Sex. Plant Reprod.*, 1996, **9**, 65.
8. A. Katsiotis, T. Schmidt and J.S. Heslop-Harrison, *Genome*, 1996, **39**, 410.
9. A. Brandes, J.S. Heslop-Harrison, A. Kamm, S. Kubis, R.L. Doudrick and T. Scmidt, *Mol. Gen. Genet.*, 1997, (in press).

SESSION II
THE HUNT FOR QTLS IN MAIZE AND
SORGHUM BREEDING

25 Years of Searching and Manipulating QTLs in Maize

Charles W. Stuber

US DEPARTMENT OF AGRICULTURE, AGRICULTURAL RESEARCH SERVICE, DEPARTMENT OF GENETICS, NORTH CAROLINA STATE UNIVERSITY, RALEIGH, NC 27965-7614, USA

1 INTRODUCTION

Research conducted early in my career focused on the study of quantitative trait variation in maize and was directed largely on attempts to measure the relative effects of epistasis in comparison with additive and dominance effects, not only for predicting population improvement but also as a component in heterosis or hybrid vigor. The inability to control many of the environmental fluctuations led me to look for some type of trait (or traits) that was correlated with the traits measured in the field, but could be measured in the laboratory, and thus would be more stable than most agronomic traits.

Twenty-five years ago, the laboratory traits that seemed most amenable for this purpose were isozymes, electrophoretic variants of enzymes. In our first attempt to relate isozymes to quantitative traits, such as grain yield in maize, allelic frequency changes at four isozyme loci were monitored over several cycles of recurrent selection for increased yield.[1] We did detect a few significant associations between isozyme marker loci and grain yield; however the evidence was not overwhelming.

The few encouraging results cited above provided evidence suggesting that the use of isozyme loci as markers for studying quantitative traits in maize might be a viable approach, however, in the early 1970s very few isozyme loci had been mapped in maize. Also, the electrophoretic technology was not developed for efficiently characterizing numerous isozyme loci on the large populations of plants required for quantitative inheritance studies. Dr. M. M. Goodman and I initiated a very productive collaboration that has resulted in the mapping of more that 40 isozyme loci using techniques whereby several enzyme systems can be characterized on a single starch gel.[2-4] My research has focused on the study of the genetic basis of phenomena such as heterosis and genotype by environment interaction, and on the use of marker technology for enhancing plant breeding efficiency in maize as well as other crops. Dr. Goodman's interests in the use of this technology have focused largely on evolutionary studies in maize and the characterization of genetic relationships among racial collections of maize.

2 EARLY MARKER STUDIES IN MAIZE

Several pioneering studies were conducted in maize during the 1970s and early 1980s that focused on associating marker genotypes with quantitative trait performances.[5-7] In several of these studies, frequency changes of alleles at a large number of marker loci were monitored over succeeding cycles of long-term selection in several populations of maize.[8-11] Allelic frequency changes at numerous loci were shown to be highly correlated with changes in several reproductive and morphological traits in maize, including the selected trait, grain yield. The impetus for the more recent activity in the use of genetic markers (isozymes and DNA-based markers) for identifying and mapping QTLs was provided by these investigations.

Several laboratories conducted investigations in which marker (isozyme or RFLP) diversity of inbred lines was correlated with performance (usually grain yield) in single-cross hybrids.[6-7] The major objective of these studies was to evaluate the use of markers for prediction of hybrid performance from crosses among untested inbred lines. The range in numbers of markers used in these studies varied from fewer than 11 isozymes to 230 RFLPs. These investigations showed that genetic distances based on marker data agreed well with pedigree data for assigning lines to heterotic groups.[12-16] However, from those studies that included field evaluations, it was generally concluded that isozyme and RFLP genotypic data were of limited usefulness for predicting the heterotic performance between unrelated inbred maize lines.[6-7]

3 EARLY STUDIES ON QTL MAPPING IN MAIZE

Results from our marker-facilitated research program at Raleigh, North Carolina, have included the identification and mapping of QTLs associated with numerous quantitative traits in 15 F_2 populations derived from seven elite inbred lines and five inbred lines with a partial exotic component (Latin American, expected to be 50%). The early studies used only isozymes as markers; however, some studies used both isozymes and RFLPs.[17-20] Although the primary focus has been on grain yield, measurements recorded on individual plants in the field evaluations have included dimensions, weights, and counts of numerous vegetative and reproductive plant parts as well as silking and pollen shedding dates. In the studies of (CO159 X Tx303)F_2 and (T232 X CM37)F_2, nearly 1900 plants were genotyped and evaluated for more than 80 quantitative traits in each population. Results from these F_2 investigations showed that QTLs affecting grain yield, and most of the other quantitative traits, were generally distributed throughout the genome; however, some chromosomal regions tended to contribute greater effects than others to trait expression.

Magnitudes of effects associated with specific QTLs have varied greatly among documented investigations. In the study of the (CO159 X Tx303)F_2 and (T232 X CM37)F_2 populations, the number of plants measured in each population (1776 and 1930) was large enough to detect factors contributing as little as 0.2% of the phenotypic variation for several yield-related traits.[17-18] Nearly 35% of the estimated genetic variation for grain yield was attributed to a major QTL in the vicinity of the isozyme marker *Amp3* on chromosome 5 in the population generated from the cross of maize line B73 with Mo17.[21]

4 MAPPING QTLs AFFECTING HETEROSIS IN THE CROSS OF B73 WITH MO17

Although QTL mapping, studies, and even some QTL manipulating studies, were conducted in a number of populations, many of the parental lines chosen for the earlier studies did not represent the elite materials that were being used in commercial maize production. In our initial studies, parental lines were chosen because they differed for many reproductive and morphological traits (we wished to maximize phenotypic variability) and because they differed for alleles at the isozyme marker loci that we were able to assay (we wished to maximize the number of polymorphic marker loci). Because of these choices of research materials, there were many skeptics who were not convinced that marker-facilitated technology had utility in a commercial breeding company. Therefore, we decided to focus on a hybrid, B73 X Mo17, which had been widely used for several years in commercial production, and even today, derivatives of the parental lines are still widely used. During the past nine years, much of our research has been based on materials generated from B73 and Mo17 or developed in combination with these lines.

The first study based on the inbred lines B73 and Mo17 was designed to explore the genetic bases of two important phenomena, heterosis and genotype-by-environment interaction. Seventy-six marker loci (9 isozyme and 67 RFLP, which were linked to about 95% of the genome), allowed us to map QTLs contributing to heterosis for grain yield in 264 backcross families.[21] These yield QTLs were located on 9 of the 10 maize chromosomes. For the mapping analyses, backcrosses to B73 were analysed separately from the backcrosses to Mo17, and for those QTLs associated with grain yield, the heterozygotes showed a higher phenotypic value than the respective homozygotes (with only one exception). These data suggested not only overdominant gene action (most likely pseudo-overdominance), but also that the detected QTLs contributed significantly to the expression of heterosis in the B73 X Mo17 hybrid. Field evaluations were made in six diverse environments (four in North Carolina, one in Iowa, and one in Illinois); however, there was little evidence for QTL-by-environment interaction.

A QTL mapping study in a population derived from the cross of B73 with Mo17 also has been reported by Beavis et al.[22] In comparison with our results[21] there were few grain yield QTLs in common. Although there were several confounding factors affecting the comparison, the most likely explanation for the differences is that both studies were based on an inadequate number of independently sampled lines. However, the low error rate in both studies indicates that each experiment identified a set of significant yield QTLs and that the number of QTLs is likely to be large.

A re-analysis of the data from the above study[21] has been completed recently using a modification of the North Carolina experimental design III.[23] The latter analysis agrees with the earlier report and strongly suggests the presence of multiple linked QTLs on most chromosomes that have significant effects on grain yield. However, the results differ from the earlier report in that the design III analysis (a very powerful tool for evaluating dominance variation) favors the hypothesis of dominance of favorable genes to explain the observed heterosis. Overdominance could not be ruled out.

4.1 Fine-mapping of QTLs in the cross of B73 with Mo17

Reduction of the size of regions identified as containing major QTLs through 'fine-mapping' is an important requirement for marker-assisted selection (MAS) and has been envisioned as the initial step for identifying single genes that ultimately could be manipulated using transformation (recombinant DNA) technology.[6] In the mapping study based on the cross of B73 X Mo17 discussed above, a region near the isozyme marker

Amp3 on chromosome 5 accounted for about 20% of the phenotypic variation for grain yield. This one large region was dissected into at least two smaller QTLs.[24] Effects at each of these two QTLs appear to act in a dominant manner, each showing significance in one testcross used in the study but not the other. These results indicate that the genetic factors are in repulsion phase linkage, thus supporting the dominance theory of heterosis. An additional segment in the targeted region on chromosome 5 showed a significant association with yield, but the effect was not consistently expressed and may be spurious. The largest of the three segments was mapped to a 13.8 cM interval near the marker *Amp3*.[24]

4.2 Mapping in the Cross of B73 with Mo17 in Stress Environments

There is little evidence for QTL by environment interaction in most investigations of traits such as grain yield in maize; however, these studies usually have been conducted under near optimum growing conditions. Such studies provide the breeder with relatively little information for making decisions on the selection of appropriate environments for evaluating QTLs that might enhance breeding strategies for growing conditions that are less than optimal. We have conducted a major investigation designed to evaluate and compare several aspects of QTL mapping under both optimum and stress environments.

Experimental materials used for this study were derived from the set of F_3 lines used in the heterosis study reported by Stuber et al.[21] From these F_3 lines, 208 F_6 recombinant inbred (RI) lines were developed by single seed descent. The lines were genotyped with about 110 RFLP and 8 isozyme markers and were backcrossed to the two parental lines, B73 and Mo17. The progenies from these backcrosses have been evaluated in four field environments (two years and two locations) each with a 2 X 2 X 2 factorial arrangement. The factors were low and high soil moisture (severe drought stress and no moisture stress), low and high planting density (36,000 plants per hectare and 72,000 plants per hectare) and low and high soil nitrogen (about 56 kilograms per hectare and 224 kilograms per hectare).

Results from the RI line backcrosses evaluated under non-stress environmental conditions were compared with earlier results using F_3 line backcrosses evaluated under similar field conditions. The comparisons showed that QTL mapping did not differ in the two generations of lines generated from the B73 X Mo17 cross. When comparing results from the RI line backcrosses evaluated in individual high stress environments with those evaluated in non-stress environments, there were nearly 10-fold differences in grain yield. However, the results indicated that no new QTLs (affecting traits such as grain yield) were detected under such high stress conditions (unpub. data). These preliminary analyses of the data imply that breeders and geneticists can rely on mapping data from favorable environments for breeding for materials adapted to stress environments.

5 MARKER FACILITATED ENHANCEMENT OF THE B73 X Mo17 HYBRID

Previous mapping studies in our program suggested that two elite inbred lines, Tx303 and Oh43, contained genetic factors that might contribute to the heterotic response of the B73 X Mo17 single-cross hybrid. Therefore, when the study cited above[21] in the population generated from the B73 X Mo17 cross was conducted, a companion study was made to identify and locate desired genetic factors in Tx303 and Oh43. Planned mean comparisons among backcrosses and testcrosses in the two studies were used to identify six chromosomal segments in Tx303 that (if transferred into B73) would be expected to enhance the B73 X Mo17 hybrid response for grain yield. Likewise, another six segments were identified in Oh43 for transfer into Mo17 that would also be expected to enhance the B73 X Mo17 hybrid response.[25] Three backcross generations were used for transfer of the identified segments into the target lines, B73 and Mo17. BC_2 families were analysed for marker genotypes which were then used to select individuals for the third backcrosses. Individuals were selected if they had the desired marker genotype in the vicinity of the donor segment to be transferred and the recipient line's genotype in the remainder of the genome. At least one marker was assayed on each of the 20 chromosome arms, and usually two markers were genotyped in the vicinity of the donor segment being transferred. Two selfing generations followed the third backcross. At each backcrossing and selfing stage, marker genotyping of plants was conducted in the same manner as for the BC_2 families. However, if a marker locus became fixed in a line, that marker was not evaluated in that line in succeeding generations and only segregating loci were analysed.

After the second selfing generation, 141 BC_3S_2 modified B73 lines were identified for testcrossing to the original Mo17. Likewise, 116 BC_3S_2 modified Mo17 lines were targeted for testcrossing to the original B73. These 257 testcross hybrids were evaluated in replicated field plots in three locations in North Carolina.[26] Forty-five (32%) of the modified B73 X original Mo17 testcrosses yielded more than the check hybrid (normal B73 X normal Mo17) by at least one standard deviation. Only 15 (11%) yielded less than the check. Evaluations of modified Mo17 X original B73 testcrosses showed that 51 (44%) yielded more grain than the normal check hybrid and only 10 (<9%) yielded less than the check hybrid. The highest yielding hybrids exceeded the check by 8 to 11% (560 to 750 kg ha^{-1}).

Based on these initial evaluations, the better performing modified lines were selected for intercrossing and were designated as "enhanced" lines. Fifteen "enhanced" B73 lines were chosen for crossing with 18 "enhanced" Mo17 lines producing 93 hybrids that were evaluated in replicated field trials in North Carolina in 1993.[25] At a planting density of about 72,000 plants per hectare, only 15 of the test hybrids produced less grain than either of two check hybrids (normal B73 X normal Mo17 and a high yielding commercial hybrid, Pioneer 3165). Six exceeded the checks by two standard deviations or more. The two highest yielding "enhanced" B73 X "enhanced" Mo17 hybrids exceeded the checks by more than 15% (1400 to 1500 kg ha^{-1}).

These "enhanced" B73 X "enhanced" Mo17 hybrids were evaluated again in 1994.[25] Twenty of these hybrids exceeded the check hybrids (same as in 1993) by two standard deviations or more, with the three best exceeding the checks by 12 to 15% (1100 to 1500 kg ha^{-1}). Although the rankings changed slightly over the two years of testing, the average yields over both years for the best yielding hybrids exceeded the check hybrids by 8 to 10% (625 to 1000 kg ha^{-1}). More importantly, the parental lines that showed superior general combining ability in 1993 also showed similar performance in 1994.

Evaluations of the highest yielding "enhanced" B73 X "enhanced" Mo17 hybrids were again conducted in 1995 in North Carolina and results corroborated those of the previous two years. In addition, nine of these hybrids were evaluated at four locations in central Iowa by Pioneer Hi-Bred International. Three of the nine exceeded the grain yields of the best Pioneer commercial check hybrids by 400 to 625 kg ha^{-1}. These results were surprising because all of the selections were based on evaluations in North Carolina.

Results from the introgression of the targeted segments from Tx303 into B73 and from Oh43 into Mo17 have demonstrated that marker-facilitated backcrossing can be successfully employed to manipulate and improve complexly inherited traits such as grain yield in maize. Not all of the six targeted segments have been successfully transferred into a single modified B73 or modified Mo17 line. There appears to be some indication that there may be no advantage in transferring more than two to four segments.

6 BREEDING SCHEME USING NEAR-ISOGENIC LINES (NILS)

The results discussed above showed that the enhancement of lines B73 and Mo17 was successful; however, the procedure for development of the "enhanced" lines (NILs) was very inefficient and is not recommended for a practical breeding program. That procedure required that the targeted segments (containing the putative QTLs) be identified in the donor lines prior to transfer to the recipient lines. In our maize program at Raleigh, we have outlined, and have tested, a marker-based breeding scheme for systematically generating superior lines without any prior identification of QTLs in the donor source(s). Identification and mapping of QTLs in the donor is a bonus derived from the evaluation of the NILs generated. Choice of the donor usually will be based on prior knowledge of its potential for providing favorable genetic factors, and, in maize, may involve appropriate heterotic relationships.

The procedure involves the development of a series of NILs by sequentially replacing segments of an elite line (the recipient genome) with corresponding segments from the donor genome. The objective is for the group of NILs generated to contain the complete genome of the donor source, with each NIL containing a different segment from the donor. Marker-facilitated backcrossing, followed by marker-facilitated selfing to fix the introgressed segments, is used to monitor the targeted segments from the donor and to recover the recipient genotype in the remainder of the genome. The number of backcrosses required will depend on the number of evaluations that can be made in the marker laboratory.

In maize, the NILs are then crossed to an appropriate tester(s) to create hybrid testcross progeny that are evaluated in replicated field trials (with appropriate checks) for the desired traits. The superior performing testcrosses will be presumed to have received donor segments that contain favorable QTLs. Therefore, QTLs are mapped by function, which should be an excellent criterion for QTL detection. The breeding scheme not only creates "enhanced" lines that are essentially identical to the original elite line, but it also provides for the identification and mapping of QTLs as a bonus with no additional cost. Obviously, the scheme is based on having a reasonably good marker map with alternate alleles in the donor and recipient lines.

This breeding strategy should be an excellent procedure for tapping into the potential of exotic germplasm. Furbeck[27] used this procedure to develop a small sample of NILs using the elite line, Mo17, as the recipient. An exotic population, derived from the accessions Cristal (MGIII), a Brazilian racial collection, and Arizona (AYA41), a Peruvian collection, was used as the donor. Using this procedure, positive segments from the exotic (such as a segment associated with Phi1 on chromosome 1) are immediately available in the adapted background (Mo17) for further breeding use. Also, and equally important, segments with negative effects (such as those associated with Dia1 and Dia2, both -595 kg ha^{-1}) are eliminated. Moreover, because each segment was incorporated independently, the detection of positive effects is not biased by negative adjacent segments (e.g., Phi1, +660 kg ha^{-1}, and Dia2, -595 kg ha^{-1}, on chromosome 1). It should be noted that Eshed and Zamir[28] have effectively used this breeding procedure to extract favorable genetic factors from a wild species of tomato.

7 CONCLUSIONS

It can be concluded that molecular-marker technology is effective for identifying and mapping QTLs in maize, as well as in many other plant species, and for studying phenomena such as heterosis and genotype by environment interaction. Positive results from marker-facilitated selection and introgression studies should encourage the use of this technology by commercial breeders for transferring desired genes between breeding lines. Appropriate use of markers should increase the precision and efficiency of plant breeding, as well as expedite the acquisition of favorable genes from exotic populations or from wild species.

The high degree of homology and synteny between sorghum and maize genomes should greatly enhance the efficiency of mapping quantitative traits in both species.[29] Also, comparative mapping with other monocots has demonstrated many examples of conserved gene order and functions,[30-32] which should prove very beneficial in the identification and mapping of useful genes in maize.

REFERENCES

1. C. W. Stuber and R. H. Moll, *Crop Sci.*, 1972, **12**, 337.

2. M. M. Goodman, C. W. Stuber, K. Newton and H. H. Weissinger, *Genetics*, 1980, **96**, 697.

3. M. M. Goodman and C. W. Stuber, 'Isozymes in Plant Genetics and Breeding, Part B', S. D. Tanksley and T. J. Orton Eds., Elsevier Scientific, Amsterdam, 1983, p. 1.

4. C. W. Stuber, J. F.Wendel, M. M. Goodman and J. S. C. Smith, 'Techniques and scoring procedures for starch gel electrophoresis of enzymes from maize (*Zea mays* L.)', *N.C. Agric. Res. Serv., N. C. State Univ., Tech. Bull*, No. 286, 1988, 87 pp.

5. C. W. Stuber, 'Plant Breeding Reviews', J. Janick Ed., John Wiley & Sons, 1992, Vol. 9, Chapter 3, p.37.

6. C. W. Stuber, 'DNA-Based Markers in Plants', R. Phillips and I. Vasil Eds., Kluwer Academic Publishers, 1994, Chapter 5, p. 97.

7. C. W. Stuber, 'Plant Breeding Reviews', J. Janick Ed. John Wiley & Sons, 1994, Vol. 12, Chapter 8, p. 227.

8. C. W. Stuber, R. H. Moll, M. M. Goodman, H. E. Schaffer and B. S. Weir, *Genetics,*1980, **95**, 225.

9. L. M. Pollak, C. O. Gardner and A. M. Parkhurst, *Crop Sci.*, 1984, **24**, 1174.

10. A. L. Kahler, *Proc. 40th Annual Corn and Sorghum Industry Research Conf.*, *American Seed Trade Assoc.*, 1985, **40**, 66.

11. R. A. Guse, J. G. Coors, P. N. Drolsom and W. F. Tracy, *Theor. Appl. Genet.,1988*, **76**, 398.

12. A. E. Melchinger, M. Lee, K. R. Lamkey, A. R. Hallauer and W. L. Woodman, *Theor. Appl. Genet.,*1990, **80**, 488.

13. A. E. Melchinger, M. Lee, K. R. Lamkey and W. L. Woodman, *Crop Sci.*, 1990, **30**, 1033.

14. E. B. Godshalk, M. Lee and K. R. Lamkey, *Theor. Appl. Genet.*, 1990, **80**, 273.

15. C. Livini, P. Ajmone-Marsan, A. E. Melchinger, M. M. Messmer and M. Motto, *Theor. Appl. Genet.*, 1992, **84**, 17.

16. B. E. Zehr, J. W. Dudley and J. Chojecki, *Theor. Appl. Genet.*, 1992, **84**, 704.

17. M. D. Edwards, C. W. Stuber and J. F. Wendel, *Genetics*, 1987, **116**, 113.

18. C. W. Stuber, M. D. Edwards and J. F. Wendel, *Crop Sci.*, 1987, **27**, 639.

19. B. S. B. Abler, M. D. Edwards and C. W. Stuber, *Crop Sci.*, 1991, **31**, 267.

20. M. D. Edwards, T. Helentjaris, S. Wright and C. W. Stuber, *Theor. Appl. Genet.*, 1992, **83**, 765.

21. C. W. Stuber, S. E. Lincoln, D. W. Wolff, T. Helentjaris and E. S. Lander, *Genetics*, 1992, **132**, 823.

22. W. D. Beavis, O. S. Smith, D. Grant and R. Fincher, *Crop Sci.*, 1994, **34**, 882.

23. C. C. Cockerham and Z-B. Zeng, *Genetics*, 1996, **43**, 1437.

24. G. I. Graham, D. W. Wolff and C. W. Stuber, *Crop Sci.*, 1997, **37**, (in press).

25. C. W. Stuber, *Proc. 49th Annual Corn and Sorghum Industry Research Conf.*, *American Seed Trade Assoc.*, 1994, **49**, 232.

26. C. W. Stuber and P. H. Sisco, *Proc. 46th Annual Corn and Sorghum Industry Research Conf.*, *American Seed Trade Assoc.*, 1991, **46**, 104.

27. S. M. Furbeck, PhD thesis, North Carolina State University, Raleigh, (Diss. Abstr. DA9330287), 1993.

28. Y. Eshed and D. Zamir, *Genetics,* 1995, **141**, 1147.

29. R. Whitkus, J. Doebley and M. Lee, *Genetics,* 1992, **132**, 119.

30. S. Ahn, J. A. Anderson, M. E. Sorrells and S. D. Tanksley, *Mol. Gen. Genet.,* 1993, **241**, 483.

31. S. Ahn and S. D. Tanksley, *Proc.Natl. Acad. Sci. USA,* 1993, **90**, 7980.

32. Y-R. Lin, K. F. Schertz and A. H. Paterson, *Genetics,* 1995, **141**, 391.

Molecular Marker Methods to Dissect Drought Resistance in Maize

S. A. Quarrie,[1] V. Lazić-Jančić,[2] M. Ivanović,[2] S. Pekić,[3] A. Heyl,[1] P. Landi,[4] C. Lebreton[1] and A. Steed[1]

[1] JOHN INNES CENTRE, NORWICH RESEARCH PARK, COLNEY, NORWICH NR4 7UH, UK
[2] MAIZE RESEARCH INSTITUTE, SLOBODANA BAJIĆA 1, ZEMUN POLJE, 11081 BELGRADE-ZEMUN, YUGOSLAVIA
[3] FACULTY OF AGRICULTURE, UNIVERSITY OF BELGRADE, NEMANJINA 6, 11081 BELGRADE-ZEMUN, YUGOSLAVIA
[4] DEPARTMENT OF AGRONOMY, UNIVERSITY OF BOLOGNA, VIA FILIPPO RE 6, 40126 BOLOGNA, ITALY

1 INTRODUCTION

The advent of molecular marker methods has revolutionized the genetic analysis of traits of agronomic importance. Thus, it is now possible to dissect in detail the mechanisms regulating drought resistance in maize. Molecular maps can be used to locate genes (or quantitative trait loci, QTLs) having major effects on drought resistance, and the relative importance of individual drought response traits in determining yield under drought can be tested from coincidence of QTLs for yield under drought and QTLs for the other traits. Alternatively, molecular markers can be used to examine allele frequency in genotypes derived from a single genetic source selected to have contrasting phenotypes (bulk segregant analysis, BSA). Differences in allele frequency between the genotypes are likely for markers which are closely associated with genes determining differences between the phenotypes, such as drought resistance.

In response to drought stress, plants accumulate the stress hormone abscisic acid (ABA) in their tissues and application of ABA to plants causes changes to the phenotype which, in many cases, mimic the effects of drought stress and which tend to improve the plant's capacity to survive water shortage. We have used the first of these molecular marker methods (QTL analysis) to examine the consequences for drought resistance in maize of genetic variation in ABA production in a population of F_2 plants and F_3 families from a cross between parents differing markedly in yield under drought (Polj17, drought resistant and F-2, drought sensitive).

Two composite populations of maize have been developed by CIMMYT (Tuxpeño Sequia, TS and Drought Tolerant Population, DTP) which have been selected for several generations on the basis of yield under drought stress in the field. After eight cycles of selection, yield of the TS population selected under severe drought conditions outyielded that of the unselected population by about 50%[1]. The yield differences between the selected and unselected DTP populations have not yet been quantified, but the DTP population has been selected further by the Maize Research Institute in Zambia for one generation under severe drought. These genetic stocks form the basis of our BSA studies to locate genes regulating drought resistance in maize. By comparing allele frequencies in the unselected and selected populations with molecular markers, it is possible to locate regions of the genome carrying genes determining drought resistance as the frequency of beneficial alleles and any linked markers would increase at loci important in determining drought resistance.

The work presented here demonstrates the utility of QTL analysis and BSA methods to test the effect of ABA accumulation on drought resistance and to identify regions of the genome important in regulating drought resistance in the crop and traits likely to be associated with that resistance. The markers closely-linked with QTLs for improved yield under drought can be used for marker-assisted selection methods to combine genes for improved drought resistance in improved varieties.

2 MATERIALS AND METHODS

2.1 QTL Analysis of a Mapping Population

A population of 81 F_2 plants and F_3 families derived from them was prepared from the cross Polj17 (high leaf ABA content) x F-2 (low leaf ABA content). The two parents also differed for many other traits likely to affect productivity under drought stress. The F_3 families were trialled under drought stress and well-watered conditions under a plastic rain-out shelter in Norwich and in the field near Bologna, Italy and near Belgrade, Yugoslavia. Anthesis-silking interval (ASI), nodal root number at maturity and grain yield were recorded in the plants. The third leaf down was also sampled from plants at around flowering time for ABA content, measured by radioimmunoassay.

The F_2 plants were genotyped with about 100 RFLP markers using a standard protocol based on Sharp et al.[2]. Genetic linkage groups for the population were constructed using the software package Mapmaker v 3.0[3].

2.2 BSA of Two Sets of Composite Populations

Two sets of composite populations were used for this work:
- Tuxpeño Sequia populations from cycle 0 (no selection for yield under drought) and cycle 8 (selected for yield under drought). Under severe drought, the yield of cycle 8 (C8) was 50% greater than that of cycle 0 (C0)[1].
- DTP populations from cycle 2 (not selected for yield under drought) and cycle 9' (after selection for yield under severe drought stress in Zambia).

To examine allele frequencies, seedlings of each population were raised and equal weights of 50 plants combined and extracted using a CTAB buffer to give DNA bulks. In addition, DNA was extracted separately from each of 36 plants from each population of Tuxpeño Sequia, cycle 0 and cycle 8. After restriction with the enzymes EcoRI, EcoRV, DraI and BamHI, DNA samples were resolved on agarose gels and Southern blotted.

Restricted DNA was hybridized with ^{32}P-labelled gDNA and cDNA probes using standard procedures[2]. Most probes came from collections of the University of Missouri (umc, npi and bnl) and California State University (csu). In addition, several drought-induced cDNAs were tested (hhu), which came from the University of Düsseldorf (the gift of Dr Michael Höfer). After hybridization and washing, filters were developed either on film or a Phosphorimager.

3 RESULTS

3.1 QTL Analysis of a Mapping Population

Although drought stress is known to stimulate the accumulation of ABA, and ABA has many effects on plant growth and development, there has been little evidence so far (except in the case of ABA-deficient mutants) that the capacity of a plant to accumulate ABA has any physiological consequence for plant productivity under drought in the field. The range of leaf ABA contents amongst our F_3 families was over five-fold in the Norwich and Belgrade trials and over two-fold in Italy. However, despite these large differences, QTL analysis of leaf ABA content in the F_3 populations provided clear evidence that, under the relatively mild drought conditions of these field trials (for example, irrigation increased mean yields in Belgrade by only 11%), variation in leaf ABA concentration had no significant effect on grain yields.

Highly significant QTLs for leaf ABA content were present on chromosomes 2 and 3 (Figure 1) in each trial, with minor effects (not stable across trials) also on chromosomes 1, 3, 5 and 9. These results confirm the major QTLs for leaf ABA content found previously in F_2 plants on chromosomes 2 and 3[4]. In contrast, QTLs for yield in each trial were clustered mainly on chromosome 8, with other significant effects under either droughted, irrigated or

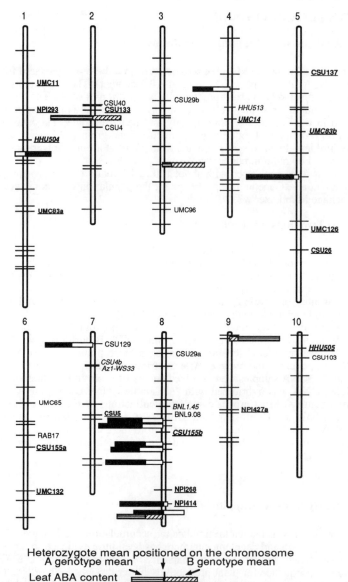

Figure 1 *Genetic map of maize chromosomes with short lines showing the positions of mapped markers. Underlined markers showed differences in allele frequency between the two TS populations. Markers in italics were positioned by reference to other maize maps. Bars are at loci having significant association with QTLs for leaf ABA content and yield. Relative differences in ABA content and yield between genotypes at each QTL are shown by the length of the bars, with mean values increasing from left to right. The heterozygote value is positioned at mid-chromosome.*

both conditions found on every chromosome except 10. In no case were QTLs for ABA and yield coincident, except for a minor ABA and yield QTL distal on chromosome 8 (Figure 1), where the heterozygous allele had the highest ABA content but an intermediate yield.

Although we found no evidence for an effect of leaf ABA content on the yield of these plants, we found several highly significant QTLs for ASI in this population located mainly on chromosome 8 in the regions of the significant effects on yields (data not presented), and this would be predicted from the phenotypic correlation between ASI and yield found by others (eg. [5]). This region of chromosome 8 also carried major genes regulating nodal root number in our population. Therefore, it is possible that the yield of these plants was also dependent at least partly on nodal root number, or some other factor that affected both root number and flowering behaviour independently, eg. plant vigour.

3.2 BSA of Two Sets of Composite Populations

In total, 28 RFLP probes were tested with the two sets of CIMMYT populations (Table 1). The map locations of these markers are shown in Figure 1. Marker locations are based on a map previously prepared by us ([4] and unpublished data). Marker loci shown in italics were located by reference to other published and unpublished maize maps. Because much of the maize genome is duplicated, about half the probes gave two bands on filters with inbred maize lines (Table 1). Map locations for the two copies were usually available.

Changes in allele frequency were found in both the TS and DTP populations (Figure 2). By comparing the positions of specific alleles with those for the parents of a mapping population, it was usually possible to infer chromosomal locations of alleles of probes that

Table 1 *Details of the RFLP Probes and Restriction Enzyme Combinations used with the CIMMYT Populations and Probe Copy Number, Identified from Hybridizations with Inbred Lines.*

Probe name	Copy no.	Restriction enzymes	Probe name	Copy no.	Restriction enzymes
bnl1.45	2	DraI	hhu513	1	All four
bn19.08	2	EcoRV	hhu(az1ws33)	2	All four
csu4	2	EcoRV	hhu(az1ws59a)	1	All four
csu5	2	All four	npi268	2	EcoRI
csu26	2	EcoRI	npi414	1	DraI
csu29	2	EcoRV	npi427	2	EcoRV
csu40	1	EcoRI, DraI	rab16	-	poor hybridization
csu103	1	EcoRI	rab17	1	All four
csu129	1	All four	umc11	2	All four
csu133	1	EcoRI	umc14	2	EcoRV
csu137	2	BamHI	umc83	2	All four
csu155	2	All four	umc96	1	All four
hhu504	2	All four	umc126	1	EcoRI
hhu505	1	All four	umc132	1	All four

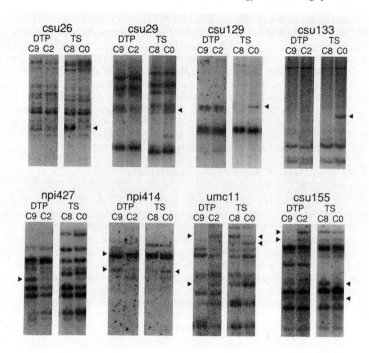

Figure 2 *Comparison of results for eight RFLP probes hybridized to restricted bulk DNA from DTP C2, DTP C9' and TS C0 and C8, showing differences between the DTP and TS populations in the alleles present and in the effects of selection for drought yield. Arrows indicate alleles showing differences in intensity between the selected and unselected populations.*

were multicopy. This occurred frequently because of the duplication of the maize genome. In particular, we found major effects on chromosomes 1, 2, 3, 5, 6, 7 and 8. Our BSA results with the TS populations confirmed the findings of a preliminary test[6] with probes that showed quantifiable differences between the TS C0 and C8 populations.

Usually, the probes showing differences in allele frequency between TS C0 and C8 also showed differences in allele frequency between DTP C2 and C9' (eg. umc11 and csu155 in Figure 2), though not always. Csu26, csu29, csu129 and csu133 showed clear differences between TS C0 and C8 but little difference between DTP C2 and C9' and vice versa for npi427. Alleles whose frequencies differed between selected and unselected populations were often not the same in the TS and DTP populations (eg. csu155, npi414 and umc11, Figure 2).

The results of hybridizations of probes with DNA from individual TS C0 and C8 plants confirmed the BSA results. Of the probes analysed so far (Table 2), all the differences in allele frequency between the selected and unselected populations were highly significant. Analysis of allele frequencies for further probes is in progress.

4 DISCUSSION

The QTL analysis of leaf ABA content identified two major QTLs on chromosomes 2 and 3 that were present in all trials and that together accounted for over 50% of the phenotypic variation. Despite the large variation in ABA content amongst lines, there was no evidence from coincidence of QTLs that leaf ABA content affected the yield of the plants in any of the

Table 2 *Allele Frequencies with Maize RFLP Probes in C0 and C8 Populations of Tuxpeño Sequia. Thirty-six Individuals were Screened from each Population.*

Probe	Band number	Number of C0 plants (% total)	Number of C8 plants (% total)	Chi-square
csu26	1	30 (83.3)	27 (75.0)	31.0
	2	6 (16.7)	23 (63.9)	‡***
csu129	1	0 (0)	5 (13.9)	17.8
	2	6 (16.7)	4 (11.1)	***
csu133	1	6 (16.7)	5 (13.9)	14.0
	2	12 (33.0)	0 (0)	**
	3	24 (66.7)	27 (75.0)	
	4	15 (41.7)	18 (50.0)	
csu155	1	10 (27.7)	19 (52.8)	12.1
	2	9 (25.0)	2 (5.6)	***
umc11	1	14 (38.8)	8 (22.2)	19.4
	2	14 (38.8)	8 (22.2)	***
	3	6 (16.6)	12 (33.3)	
	4	14 (38.8)	27 (75.0)	

‡ ** and *** indicate significance of chi-square at $P<0.01$ and 0.001 respectively.

trials. This result confirms previous findings with S_1 families from high-ABA and low-ABA composites derived from the same cross (Polj17 x F-2)[7]. Here, a small decline in yield was associated with very high leaf ABA concentrations. We are currently studying the association between leaf ABA concentration and grain yield under more severe drought using immortalized F_3 families from this cross.

BSA with the two sets of composites identified several regions of the maize genome where significant differences in allele frequency had taken place during selection for high drought yield. Many of the probes showing major differences in allele frequencies in the TS populations also showed differences in the DTP populations but this was not always so. It is likely, therefore, that yield under drought is regulated by several genes that differ between the TS and DTP populations.

Among the RFLP probes tested were several from the University of Düsseldorf (hhu) which were cDNAs known to be induced in maize leaves by drought stress. However, only two of these cDNAs (coding for two members of a family of proline-rich proteins) showed any clear evidence for differences in allele frequencies between the TS C0 and C8 populations. It is unlikely, therefore, that the majority of these drought-induced genes have any significant effect on the yield of maize under drought stress.

Recent work by Agrama and Moussa[8] and Ribaut et al.[9,10] has located regions of the maize genome carrying major genes for both yield under drought and ASI (an important factor in determining drought yield). Several of these genes were on chromosomes 1, 5, 6 and 8 close to markers we showed to differ in allele frequency (specifically, umc11, umc83a, csu26, csu155a, b and bnl9.08). Thus, our results show the utility of the BSA method to identify regions of the genome important in regulating drought resistance in maize. BSA is also useful for testing cDNAs as probes for candidate genes regulating drought resistance. Furthermore, the alleles specific for improved yield can be used for marker-assisted selection methods to combine genes for improved drought resistance in improved varieties.

Acknowledgement

We are grateful to the Royal Society for financial support for part of this work.

References

1. J. Bolaños and G. O. Edmeades, *Field Crops Res.*, 1993, **31**, 233.
2. P. J. Sharp, M. Kreis, P. R. Shewry and M. D. Gale, *Theor. Appl. Genet.*, 1988, **75**, 286.
3. E. S. Lander, P. Green, J. Abrahamson, A. Barlow, M. J. Daly, S. E. Lincoln and L. Newburg L, *Genomics*, 1987, **1**, 174.
4. C. Lebreton, V. Lazić-Jančić, A. Steed, S. Pekić and S. A. Quarrie, *J. Exp. Bot.*, 1995, **46**, 853.
5. J. Bolaños and G. O. Edmeades, *Field Crops Res.*, 1996, **48**, 65.
6. S. A. Quarrie, A. Heyl, A. Steed, C. Lebreton, V. Lazić-Jančić, 'Physical stresses in plants. Genes and their products for tolerance' S Grillo, A Leone (eds.) ESF meeting, Maratea, September 1995, Springer, Berlin, 1996, p. 141.
7. S. A. Quarrie, S. Pekić, V. Lazić-Jančić, M. Ivanović and G. Vasić, 'Proceedings 50th Anniversary, Maize Research Institute' Ž Videnović (ed), Maize Research Institute, Zemun Polje, Belgrade-Zemun, 1995, p. 121.
8. H. A. S. Agrama and M. E. Moussa, *Euphytica,* 1996, **91**, 89.
9. J.-M. Ribaut, D. Hoisington, J. A. Deutsch, C. Jiang, D. Gonzales de Leon, *Theor. Appl. Genet.*, 1996, **92**, 905.
10. J.-M. Ribaut, C. Jiang, D. Gonzalez-de-Leon, G. O. Edmeades, D. A. Hoisington, *Theor. Appl. Genet.*, 1996, (in press).

Physiological and Genetic Bases of Nitrogen Use Efficiency in Maize

P. Bertin,[1] A. Charcosset[1] and A. Gallais,[1,2]

[1] STATION DE GÉNÉTIQUE VÉGÉTALE INRA-UPS-INA PG, FERME DU MOULON, 91190 GIF-S/YVETTE, FRANCE
[2] INSTITUT NATIONAL AGRONOMIQUE DE PARIS-GRIGNON, 16 RUE CLAUDE BERNARD, 75231 PARIS CEDEX 05, FRANCE

1 INTRODUCTION

Nitrogen fertilization is necessary for non symbiotic fixing plants. Combined with genetic improvement, it has been a powerful tool for increasing yield, especially for cereals and maize. For maize, nitrogen fertilization is given in only one manuring and under some conditions it can be at the origin of pollution, mainly with silage maize where slurry is used. Taking into account the consequence of agricultural policy to maintain a sufficient "margin" for the farmer and to avoid pollution by nitrates, it is necessary to have varieties of maize which absorb and metabolize nitrogen from fertilizer well.

Genetic variability for nitrogen use efficiency has already been observed with differences in responsiveness[1,2]. It appears clearly for the two components of nitrogen use efficiency : the nitrogen uptake efficiency and the nitrogen utilization efficiency [3,4,5,6].

The aim of our study, using recombinant inbred lines from material adapted to the north of France for grain production was;
- to find criteria for nitrogen use efficiency (NUE);
- to identify and characterize some quantitative trait loci (QTL) involved in the observed genetic variation for NUE.

2 MATERIALS AND METHODS

One hundred recombinant inbred lines crossed to the same unrelated tester have been used. They derived from the cross between a french flint and early line (F2) and an iodent late line. The importance of differences in earliness is decreased by the cross with the tester.

Two N-levels were used : a normal nitrogen fertilization (N^+) with 175 kg N/ha applied at the time of sowing and no nitrogen fertilization (N^-), all the nitrogen being supplied from the soil, a supply estimated at 50 - 60 Kg N/ha.

The experimental design was a split-splot with 3 replications. The experiment was developed in 1994 and 1995.

The observed traits were the following :
    . yield and its components (kernel number/ear thousand kernel weight);
    . earliness : flowering date and moisture content;

. anthesis-silking interval;
. morphological traits : plant height, ear height;
. leaf senescence;
. nitrogen content in plant parts (leaves, stems + sheaths, grains) at flowering and harvest.

A genetic map was first developed with 74 RFLP markers, but now, more than 150 RFLP markers are available including probes of specific genes. 5 specific probes of genes involved in the nitrogen metabolism have been specially considered : 2 nitrate reductase, 2 glutamine synthetase and one GOGAT. The aim is to see their coincidence with QTL for nitrogen use efficiency.

To locate QTLs we have used Plab-QTL [7].

3 RESULTS

3.1 Genetic Variability - Heritabilities - Correlations among traits

3.1.1 Yield and its components. As expected, grain yield was strongly affected by the reduction in nitrogen fertilization with a decrease of about 40 % on average in both years (table 1). Such a decrease is mainly due to the decrease in the number of kernels per ear (32 %). However, there was also an increase in barrenness. The thousand kernel weight is reduced by only 10 %. Note that the decrease in the number of kernels per ear is due to the decrease in the number of kernels per row, the number of rows being nearly unaffected.

Table 1 *Mean, genetic and environmental variances and plot heritabilities.*

Trait	year		mean		variance		
			qx/ha	%	σ^2_g	σ^2_r	h² plot
yield	1994	N+	82.5	100	36.2	36.2	0.50
		N-	50.5	61.2	65.8	66.9	0.50
	1995	N+	79.9	100	31.2	28.9	0.52
		N-	49.1	61.5	23.9	72.1	0.25
kernel number	1994	N+	328.4	100	657.6	486.6	0.56
		N-	220.2	67.1	959.3	1005.2	0.48
	1995	N+	342.14	100	714.8	598.5	0.54
		N-	233.7	68.3	626.7	1592.5	0.28
kernel weight	1994	N+	248.9	100	252.3	39.0	0.87
		N-	227.1	91.3	227.9	65.0	0.78
	1995	N+	256.9	100	283.7	143.3	0.66
		N-	231.6	89.1	230.0	163.4	0.58

σ^2_g genetic variance ; σ^2_r : environmental variance ; h²plot = plot heritability

For yield and its components the ANOVA showed a strong genotypic effect (table 2). Genotype x nitrogen fertilization interaction was significant for grain yield and kernel number but not for thousand kernel weight. The same situation has been observed with the genotype x year interaction.

Table 2 *ANOVA for yield and its components, with the three factors: genotype, year and nitrogen fertilization.*

Factors	df	yield F[(1)]	yield Variance	kernel number F	kernel number Variance	kernel weight F	kernel weight Variance
Year	1	**		***		***	
Nitrogen	1	***		***		***	
Nitrogen x Year	1	ns		ns		*	
Genotype	100	***	18.5	***	280.4	***	216.9
Genotype x Year	100	**	8.4	***	198.4	ns	3.7
Genotype x Nitrogen	100	*	5.6	**	194.5	ns	6.4
Genotype x Nitrogen x Year	100	**	5.0	ns	37.4	ns	17.4
Residual	692[(3)]		46.3		976.8		117.3

(1) ***, **, * significance at respectively 0.1% , 1% , 5% (3) df=592 for kernel number and 596 for kernel weight

The genotype x nitrogen fertilization interaction represents only 28 % of the genetic variation. This is also illustrated by the high correlation between the two levels of fertilization in both years (0.70 in 1994, 0.78 in 1955). However, in spite of a significant interaction genotype x nitrogen fertilization x year there appears a significant phenotype correlation between responsiveness in 1994 and responsiveness in 1995 : 0.53 for yield and 0.77 for kernel number [responsiveness is defined as the difference for a trait between the two performances, with and without nitrogen fertilization].

In 1994 the phenotypic correlation between responsiveness for yield and responsiveness for kernel number per ear was very high (0.89). This demonstrates as suggested by the results on means, that kernel number per ear explains very well the responsiveness of genotypes to nitrogen. Then the correlation observed between responsiveness in 1994 and 1995 for kernel number (0.77), higher than that observed for yield, illustrates a repeatability of the genotype x nitrogen interaction.

Heritabilities for grain yield with and without nitrogen fertilization were the same in 1994 but in 1995, heritability without nitrogen was very low. In both years, environmental variances were strongly increased in the absence of nitrogen fertilization. This is due to the appearance of the soil heterogeneity which is masked with normal nitrogen fertilization. Note that in 1994 an increase in the genetic variance was also observed in the absence of nitrogen fertilization. Both years were indeed different. Spring 1994 was favourable, but there were unfavourable autumn conditions, and, inversely, spring 1995 was rather unfavourable where as summer and autumn were favourable.

As far as genetic advance due to direct and indirect selection is concerned, it appears that, for grain yield, in order to have a general adaptation, it is better to select with nitrogen than without nitrogen. Selecting with nitrogen leads to a loss of gain without nitrogen of 20 %, while selecting without nitrogen leads to a loss of gain of 40 % in presence of nitrogen. Furthermore, the response without nitrogen appears to be more erratic.

3.1.2 Traits related to nitrogen metabolism. Due to a strong negative relationship between protein-content and dry-matter yield at flowering, the heritability of protein yield appears very low, mainly in 1995 (table 3). Nevertheless, in 1994, protein yield at flowering without nitrogen fertilization was genetically correlated to the responsiveness in grain yield through an effect on the kernel number. Then in 1994, N-metabolism seems to explain a part of the genotype x nitrogen fertilization interaction observed for yield.

However, due to a stronger negative correlation between yield and N-content (perhaps due to the cold stress in the spring) this was not observed in 1995.

Table 3 *Means, variances and heritabilities for different traits related to nitrogen metabolism at flowering.*

Trait *		mean	σ_g^2	σ_e^2	h²plot
dry-matter yield (g/pl.)	N+	74.3	42.1	57.9	0.42
	N-	65.6	20.8	50.2	0.29
%N plant	N+	1.67	52 10⁻⁴	96 10⁻⁴	0.34
(g N/g dry-matter)	N-	0.93	40 10⁻⁴	64 10⁻⁴	0.38
qN (g/plant)	N+	1.24	89 10⁻⁴	222 10⁻⁴	0.29
	N-	0.60	8 10⁻⁴	85 10⁻⁴	0.09
qN stem (g/plant)	N+	0.64	20 10⁻⁴	86 10⁻⁴	0.19
	N-	0.30	ns	29 10⁻⁴	-
qN leaves (g/plant)	N+	0.59	21 10⁻⁴	46 10⁻⁴	0.32
	N-	0.31	4 10⁻⁴	21 10⁻⁴	0.16

* qN= N quantity per plant

Two other traits appear associated with susceptibility to nitrogen stress :
- leaf senescence, studied through the relative rank of the last senescent leaf : a negative correlation with yield (- 0.80) in the absence of nitrogen fertilization is observed ;
- anthesis-silking interval, with a negative correlation with yield (- 0.75), also in the absence of nitrogen fertilization.

Such results have also been observed by different authors (see this Congress). Furthermore such traits are known to be associated with any type of stress limiting growth.

3.2 Detection of QTLs

For grain yield and kernel number it appears to be easier to detect QTLs with than without nitrogen fertilization (Figure 1). QTLs detected without nitrogen fertilization tend to be a subset of those detected with nitrogen fertilization. Such a conclusion is still clearer on the average of both years 1994 and 1995 and less variation is explained by markers without nitrogen than with. This could be due to a greater experimental error, mainly in 1995 where very few QTLs are detected without nitrogen fertilization. It must be underlined, that for thousand kernel weight, a trait which does not show any interaction between genotype and level of nitrogen fertilization or with year, the detected QTLs are the same whatever the conditions.

QTLs detected for yield and its components do not coincide with those detected for earliness except for those on chromosome 1. In the presence of nitrogen fertilization, favorable alleles are found in both parents : generally F2 has more favourable alleles for thousand kernel weight and Io has more favourable alleles for kernel number. However, F2 brings some favourable alleles for kernel number as Io for kernel weight. Then by manipulating QTLs for yield and its components, it seems possible to improve yielding ability under condition of nitrogen fertilization.

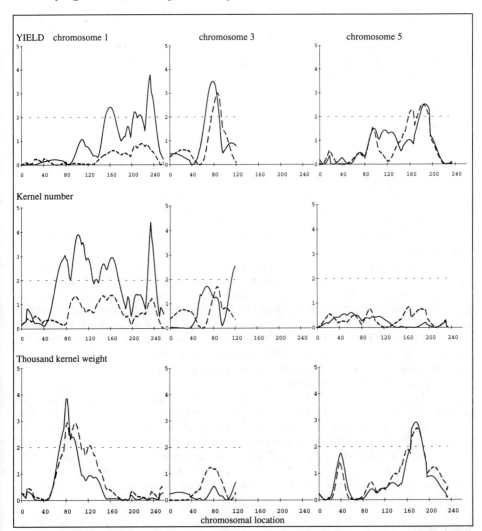

Figure 1 *Some examples of QTL detected for grain yield , kernel number and thousand kernel weight.* Graphs show LOD values according to chromosomal location and nitrogen level (N+ , N- respectively unbroken and broken graph).

A QTL for leaf senescence appears on chromosome 7 ; it explains 42 % of the genetic variance of this trait. It may correspond to a "major" gene. A QTL for responsiveness of kernel number per ear is located on chromosome 6. Unfortunately, there is no coincidence between these QTLs and those detected for yield and responsiveness for yield. They can correspond to QTL affecting yield or its responsiveness to N but be undetected at the level of yield itself, due to its complexity.

Considering the probes of specific genes involved in the nitrogen metabolism, there appear some coincidences with detected QTLs for yield or its components. On five probes, three, corresponding to Glutamine Synthetase, are in the region of QTLs for thousand kernel weight. Such a correspondence seems to have a too low probability to be due to chance. Then, this way of using probes of known genes assumed to be involved in the physiology of the plant, i.e. the candidate gene approach, could be very useful for studying the physiological and genetic meaning of the detected QTLs and for finding markers linked to the QTLs, a situation which will increase the efficiency of Marker Assisted Selection.

4 CONCLUSION

From our preliminary results and with the material studied, a genotype x nitrogen fertilization interaction appears. Undoubtedly, it could have been greater with more diverse material. This means that it is possible to develop new hybrids with a better nitrogen use efficiency. Apparently, it seems easier to improve NUE with a normal nitrogen fertilization than with a too restricted nitrogen fertilization. Our first conclusions have greater consequence for the improvement of maize in normal conditions than in stress conditions.

In the absence of nitrogen fertilization, the traits most closely associated to susceptibility to this stress are leaf senescence and anthesis-silking interval. Tollenaar and Mihajlovic[8] have already shown, that the improvement of leaf area duration is related to improvement of tolerance to various types of stress and of yield. A short anthesis-silking interval (or even protogyny) has been shown by Boyat and Robin[9] to be related to a greater nitrate reductase efficiency. Then breeding in two such criteria could be very efficient for improving nitrogen stress tolerance. Maize breeders now select regularly for stay green. More emphasis could be placed on the anthesis-silking interval.

QTLs are difficult to detect without nitrogen fertilization, they explain less variation than with nitrogen fertilization and they tend to be a subset of those detected with nitrogen fertilization. However this could be due to the lack of accuracy without nitrogen fertilization. This shows, at least, that the detected QTL under such conditions are not responsible for the genotype x nitrogen fertilization interaction. Such an interaction could be due to QTLs with a smaller effect. These preliminary results show that markers could be used with our studied material to develop lines or hybrids more efficient under condition of normal nitrogen fertilization. The adaptation to nitrogen stress by the use of markers appears more difficult.

References

1.   L.G. Balko and W.A. Russel, *Maydica*, 1980, **25**, 64-79.
2.   L.B. Balko and W.A. Russel, *Maydica*, 1980, **25**, 81-94.
3.   E.J. Kamprath, R.H. Moll and N. Rodriguez, *Agron. J.*, **74**, 955-958.
4.   E.L. Anderson, E.J. Kamprath and R.H. Moll, *Crop Sci.*, 1985, **25**, 598-602.
5.   R.H. Moll, E.J. Kamprath and W.A. Jackson. *Proc. 37th annual corn and sorghum industry research conf.*, 1983, pp. 163-175.
6.   G. Mollaretti, M. Bosio, E. Gentinetta and M. Motto, *Maydica*, 1987, **32**, 309-323.
7.   N.F. Utz and A.E. Melchinger, *J. QTL*, 1995, **1**.
8.   M. Tollenaar and M. Mihajlovic, *Crop Sci.*, 1991, **31**, 119-124.
9.   A. Boyat and P. Robin, *Ann. Amélior. Plantes*, 1977, **27**, 389-410.

Identification of QTLs for Abscisic Acid Concentration in Drought-stressed Maize and Analysis of their Association with QTLs for Other Physiological Traits

M. C. Sanguineti, S. Salvi, R. Tuberosa, P. Landi, E. Casarini, E. Frascaroli and S. Conti

DIPARTIMENTO DI AGRONOMIA, VIA FILIPPO RE 6, 40126 BOLOGNA, ITALY

1 INTRODUCTION

The increase in abscisic acid (ABA) concentration observed under conditions of water deficit represents a pivotal factor in the adaptive response of plants to drought[1]. The majority of information on the genetic basis of the accumulation of ABA in plants undergoing water stress has been obtained from studies carried out in cereals[2]. These studies indicated that leaf ABA concentration (L-ABA) in drought-stressed plants is a quantitatively-inherited trait which can influence, directly or indirectly, many other traits.

The advent of molecular marker techniques allows for a more precise identification of the quantitative trait loci (QTLs) harbouring the genes for the trait under investigation and for a more direct evaluation of the effects which one chromosomal region may have on different traits either through linkage and/or pleiotropy. Furthermore, it is possible to compare the relative importance of QTLs in different genetic backgrounds. QTLs for L-ABA have been identified in barley[3,4], wheat[5], and maize[6]. In the case of maize, up to seven QTLs were found to control L-ABA at subsequent growth stages in an F_2 population grown in a soil glasshouse.

In the present work, L-ABA and other physiological traits were analysed in F_4 families grown in the field under drought conditions in order i) to identify QTLs affecting L-ABA at different growth stages and verify the consistency of their effects during two years and ii) to investigate the overlapping between the QTLs for L-ABA and the QTLs of some physiological traits involved in the adaptive response to drought.

2 MATERIALS AND METHODS

The source population was the F_2 of the Os420 x IABO78 cross. Parental lines were chosen according to previous experiments[7] which had shown that Os420 has a L-ABA value 2- to 3-fold higher than IABO78. Following the single seed descent procedure, 80 F_4 random families were derived. These families were tested in field experiments conducted in 1994 and 1995 at Cadriano near Bologna, Italy, according to an RCBD with three replications. Plants were grown according to the field practices usually adopted for maize in our region, except for planting date and irrigation. Planting was delayed to have anthesis during the driest time of the summer in our region. Each plot included 12 plants (4.0 plants m[-2]). To achieve an adequate level of water stress, irrigation volumes corresponded to approximately 50% of the actual evapotranspiration, after accounting for rainfall. The traits recorded on the eight central plants/plot were: a) leaf ABA concentration measured, according to Hanway's classification[8], at stage 3 (L-ABA-3) and 4 (L-ABA-4) corresponding to rapid stem elongation and anthesis, respectively; the fifth and the third leaves from the top were considered at the first and second samplings, respectively; free-ABA concentration was determined on crude aqueous extracts of leaf tissue with a radioimmunoassay technique[9]; b) drought stress index visually scored at stage 3 (DSI-3) and 4 (DSI-4) on a scale from 1 (no stress) to 5 (high stress level); c) stomatal conductance measured with an open-chamber porometer at stage 4 (SC-4) on the abaxial and adaxial surfaces of two of the leaf blades sampled for L-ABA; d) leaf temperature (LT) measured with an infrared thermometer at stage 4 in 1994 (mean of two measurements) and between stages 3 and 4 in 1995 (mean of seven measurements).

DNA was digested with *Eco*RI, *Hind*III, and *Bam*HI and hybridized with 109 RFLP probes kindly provided by the University of Missouri (Columbia). The procedures utilized to reveal polymorphisms are those described in[10]. The genetic map was computed using the program JOINMAP[11] and an LOD score of 2.5. Composite interval mapping (CIM) procedure was used to identify the QTL's position and effect (PLABQTL program[12]). Cofactors were chosen by stepwise regression (F to Enter = F to Drop = 3.5). Only additive effects were included in the model. LOD scores equal to 2.0 and 1.5 were chosen to declare a putative QTL for L-ABA and the other investigated traits, respectively. The supporting interval of each QTL was comprised within LOD values one unit lower than the LOD value at the peak.

3 RESULTS AND DISCUSSION

The ANOVA showed that differences among F_4 families were significant (P 0.01) for all the investigated traits (data not shown). Significant differences (P 0.01) for L-ABA were also found between stages within years; also the "stage x progeny / year" interaction w a s

Table 1 *Mean Values, Ranges, and Heritability of the Investigated Traits*

	L-ABA-3 (ng g[-1] DW)	L-ABA-4 (ng g[-1] DW)	DSI-3 (score)	DSI-4 (score)	SC-4 (cm s[-1])	LT (°C)
1994						
Mean F_4	587	366	2.45	2.57	0.49	29.6
Min	427	225	1.00	1.33	0.21	28.5
Max	983	790	4.33	4.00	0.85	31.1
h^2	0.79	0.88	0.71	0.62	0.44	0.45
1995						
Mean F_4	327	350	2.67	2.28	0.37	31.2
Min	213	186	1.00	1.33	0.08	30.1
Max	524	648	4.50	4.00	0.70	32.9
h^2	0.60	0.88	0.86	0.88	0.76	0.84

significant. The values of L-ABA (Table 1) were within the range reported in previous studies[7,13,14] in which maize genotypes had been tested under drought conditions. The mean values of DSI, SC, and LT (Table 1) were also within the range of those reported in previous studies conducted under drought-stress conditions[13,14] thus indicating that the intensity of drought was adequate for the purpose of this study. The heritability (h^2) values for L-ABA computed on a family mean basis were quite high, especially at the second sampling.

3.1 QTL Analysis for L-ABA

The linkage map spanned 1578 cM which corresponds approximately to 75% of the length of the maize map. The map included 13 linkage groups which were all assigned to the ten chromosomes based on the information available on the map position of the UMC probes. The results of QTL analysis for L-ABA are shown in Table 2. Only QTLs whose supporting intervals did not overlap are reported. The QTLs have been listed according to their position on the map. The number of significant (LOD > 2.0) QTLs for L-ABA at each sampling varied from two to eight. Altogether, 16 QTLs showed significant effects in at

Table 2 *Additive Effect (ng ABA g^{-1} DW) and LOD Score at the QTLs for L-ABA*

Chrom.	Closest probe	1994		1994		1995		1995	
		L-ABA-3		L-ABA-4		L-ABA-3		L-ABA-4	
1	UMC11	47[(1)]	(2.97)[(2)]	-		-		-	
	UMC13	-		49	(6.62)	-		-	
2	UMC6	-		-		-		34	(2.54)
	CSU133	-		41	(5.56)	-		72	(8.88)
	CSU109a	-		33	(5.56)	34	(7.17)	32	(3.46)
3	NPI268	-		-		13	(3.64)	-	
	CSU36b	-		32	(3.11)	-		-	
	CSU25b	-28	(2.70)	-		-		-	
4	UMC31	-		16	(2.65)	18	(2.47)	32	(2.87)
5	UMC51	-		-		18	(2.83)	-	
6	UMC59	-		-		-		25	(2.01)
	UMC132	-		-		-		-25	(2.22)
7	ASG8	-		-		-21	(2.80)	-	
	CSU17b	-		-		25	(4.85)	-	
	UMC35	-		-14	(2.57)	-21	(2.89)	-	
9	BNL14.28	-		-28	(4.23)	-32	(5.01)	-	
	QTL (no.)	2		7		8		6	
	R^2 (%)	18		60		51		65	

[(1)] Computed as (Os420 - IABO78)/2
[(2)] LOD score

least one of the four samplings. The alleles increasing L-ABA were mainly contributed by Os420 (i.e. the high L-ABA parent), although in at least five cases the plus alleles were those of IABO78. Five QTLs showed significant effects in two or three samplings. At each one of these QTLs, the direction of the effect of the parental allele was always consistent across samplings. The second QTL on chromosome 2 had the strongest effect and peaked near CSU133, exactly where a QTL for L-ABA was reported in a different maize population[6]. Interestingly, this QTL was detected only at the second sampling in both years, despite the fact that in 1994 the L-ABA value was higher at the first sampling. Therefore, the action of this QTL seems independent from the level of water stress experienced by the plant when the leaf samples were collected. It is possible that the gene/s underlying this QTL is/are expressed only at a later growth stage.

The coefficient of determination (R^2) of the multiple regression varied from 18 to 65%. It is likely that other QTLs for L-ABA went undetected because i) they are located in regions not sufficiently covered by our map, ii) their absolute effect is small, and/or iii) the number of progenies herein tested was not sufficiently large.

The importance of the QTL on chromosome 2 was confirmed in another experiment in which Southern analysis was carried out on F_4 progenies obtained after two cycles of divergent selection for L-ABA starting from the same F_2 population (Os420 x IABO78)[15]. The analysis of the DNA samples from the eight progenies with the lowest and highest L-ABA values indicated very clearly the importance of the region on chromosome 2 surrounding CSU133. In fact, seven of the eight F_4 families selected for high L-ABA were homozygous for the Os420 marker allele and all the eight F_4 families selected for low L-ABA were homozygous for the IABO78 allele.

3.2   Association Among QTLs for L-ABA and QTLs for Physiological Traits

Significant QTLs were also identified for the other investigated traits (results not shown). Only those QTLs whose supporting intervals overlapped with those of QTLs for L-ABA were further considered. Table 3 reports the direction of the effect (increase or decrease) which was associated to an increase in L-ABA at the corresponding QTLs. In general, an increase in L-ABA was associated to a higher DSI, lower SC, and higher LT. These associated effects were particularly evident for the first QTL on chr. 6 at the second sampling in 1994. These results indicate a higher level of water stress of the plants with the allele increasing L-ABA. Only at the first QTL for L-ABA on chromosome 7, was the increase in L-ABA associated with higher SC and lower LT. Provided that these results are interpreted on the basis of pleiotropic effects, an increased L-ABA may have influenced favourably morpho-physiological traits with a positive effects for the water balance of the plant. Previous work has shown that an increase in ABA in drought-stressed maize plantlets has a positive effect on their rate of root growth[16]. Surprisingly, the QTL on chromosome 2 with the largest effect on L-ABA was not associated with any of the QTLs for the investigated traits. This result further supports the hypothesis that the expression of the gene/s underlying this QTL is more likely controlled by the growth stage rather than the water balance of the plant which greatly affects DSI, SC, and LT.

4 CONCLUSIVE REMARKS

QTL analysis has allowed the identification of several chromosomal regions affecting

Table 3 *Direction (↑: increase; ↓: decrease) of the Effect Observed at QTLs for Physiological Traits Overlapping the QTLs for L-ABA. The Direction Reported at Each QTL Refers to the Allele Increasing L-ABA*

Chrom	DSI-3		DSI-4		SC-4		LT	
	94	95	94	95	94	95	94	95
1	-	-	-	-	-	↓	-	-
	-	-	-	-	↓	-	↑	↑
2	-	-	-	-	-	-	↑	↑
	-	-	-	-	-	-	-	-
3	↓	-	↓	-	-	-	-	-
	-	-	-	-	-	-	-	-
	-	-	-	-	↓	-	-	-
4	-	-	-	↑	-	-	-	↑
5	-	-	-	-	-	-	-	-
6	-	-	↑	-	↓	-	↑	-
	↓	-	-	-	-	↓	-	-
7	-	-	-	-	↑	-	↓	-
	-	-	↑	-	-	-	-	-
	-	-	-	-	-	-	↓	-
9	↑	-	-	↑	-	-	↓	-

L-ABA in the investigated F_4 families. Presently, it is difficult to elucidate the physiological basis of the QTLs for L-ABA. It seems unlikely that these QTLs represent one or more of the structural genes coding for the enzymes involved in the biosynthesis of L-ABA. In fact, the supporting intervals of the QTLs herein evidenced did not include the map position of structural genes known to be involved in ABA biosynthesis in maize. Genes controlling morphological traits (e.g. root morphology) affecting the water balance of the plant could partially account for the observed results. Alternatively, some QTLs could influence ABA biosynthesis through the intensity of the transduction signal associated with turgor loss, a major determinant in the increase of ABA concentration. The QTL with the strongest effect was located on chromosome 2; its position coincided with that of a QTL for L-ABA and root-pulling strength reported in a different maize population[6].

The comparative analysis of the effects associated with the QTLs for drought stress index, stomatal conductance, and leaf temperature, whose supporting interval overlapped with the QTLs for L-ABA, in most cases did not evidence a positive role of L-ABA in sustaining a more favourable water balance. Although it is not possible to ascertain if these associated effects are due to linkage or pleiotropy, in the present study L-ABA seemed to represent mainly an indicator of the stress level experienced by the plant at sampling, rather than having the role of a causal agent capable to favourably influence the plant under drought stress in the field.

Acknowledgements

The authors wish to thank Sandra Stefanelli for skillful technical assistance, T.A. Musket of the University of Missouri (Columbia) for the RFLP probes, and S. Quarrie for providing the ABA-monoclonal antibody. Research supported by National Research Council of Italy, Special Project RAISA, Sub-project No. 2, Paper No. 2857. Contribution of Interdepartmental Centre for Biotechnology, University of Bologna.

References

1.   T.A. Mansfield, In: R.D. Davies (ed.) 'Plant Hormones and their Role in Plant Growth and Development', Martinus Nijhoff, Dordrecht, 1988, p. 411.
2.   S.A. Quarrie, In: W.J. Davies, and H.G. Jones (eds) 'Abscisic Acid Physiology and Biochemistry', Bios Scientific Publishers, Oxford, 1991, Chapter 16 p. 227.
3.   M.C. Sanguineti, R. Tuberosa, S. Stefanelli, E. Noli, T.K. Blake, and P.M. Hayes, *Russian J. Pl. Physiol.*, 1994, **41**, 572.
4.   E. Noli, M.C. Sanguineti, R. Tuberosa, and E. Schilirò, V Int. Oat Conf. and VII Int. Barley Genetics Symposium, Vol. 1, Saskatoon, Canada, 1996, p.366.
5.   S.A. Quarrie, A. Steed, C. Lebreton, M. Gulli, C. Calestani, and N. Marmiroli, *Russian J. Pl. Physiol.*, 1994, **41**, 565.
6.   C. Lebreton, V. Lazic-Jancic, A. Steed, S. Pekic, and S.A. Quarrie, *J. Exp. Bot.*, 1995, **46**, 853.
7.   S. Conti, P. Landi, M.C. Sanguineti, S. Stefanelli, and R. Tuberosa, *Euphytica*, 1994, **78**, 81.
8.   J.J. Hanway, *Agron. J.*, 1963, **55**, 487.
9.   S.A. Quarrie, P.N. Whitford, N.E.J. Appleford, T.L. Wang, S.K. Cook, and I.E. Henson, *Planta*, 1988, **173**, 330.
10. M.A. Saghai-Maroof, K.M. Soliman, R.A. Jorgesen, and R.W.Allard, Proc. Natl. Acad. Sci., USA, 1984, **81**, 8014.
11. P. Stam, *Plant J.*, 1993, **5**, 503.
12. H.F. Utz and A.E. Melchinger, *J. Quantitative Trait Loci*, 1996, http:// probe.nalusda.gov:8000/otherdocs/jqtl/2.
13. P. Landi, S. Conti, F. Gherardi, M.C. Sanguineti, and R. Tuberosa, *Maydica*, 1995, **40**, 179.
14. R. Tuberosa, M.C. Sanguineti, and P. Landi, *Crop Sci.*, 1994, **34**, 1557.
15. P. Landi, M.C. Sanguineti, S. Stefanelli, R. Tuberosa, M.M. Giuliani, and S. Conti, XVIIth Conf. 'Genetics, biotechnology and breeding of maize and sorghum', Thessaloniki, Greece, 1996, p. 118.
16. I.N. Saab, R.E. Sharp, J. Pritchard, and G.S. Voetberg, *Plant Physiol.*, 1990, **93**, 1329.

QTL Analysis of Forage Maize Traits: Comparison Across Testers

Th. Lübberstedt and A. E. Melchinger

INSTITUTE OF PLANT BREEDING, SEED SCIENCE AND POPULATION GENETICS,
UNIVERSITY OF HOHENHEIM, 70599 STUTTGART, GERMANY

1 INTRODUCTION

In temperate climates, maize (*Zea mays* L.) plays an important role as forage crop. In Germany, the acreage of grain and corn-cob-mix maize (345000 ha) was in 1994 significantly exceeded by the acreage of forage maize (1208000 ha). Forage maize is mainly used for feeding of dairy cattle and bullfattening. Hence, the major goals are to maximize the dry matter yield, increase the nutritive quality of the feed, and obtain an optimum of dry matter concentration for the silage process and for a high intake and good digestibility by ruminants.

Remarkable progress in forage maize breeding was gained during the last few decades by conventional hybrid breeding. The development of a large number of molecular markers for maize raises the question, whether their application can further enhance the efficiency of forage maize breeding. One important application of markers is the prediction of the best hybrid within a large set of possible line combinations based on the genetic distances of the parent lines. However, at least for line combinations between different heterotic pools of European material, this was not possible[1]. Another important application is marker assisted selection (MAS) for complex traits like dry matter yield or digestibility of the fodder. Here, the first step is to identify molecular markers closely linked to quantitative trait loci (QTL). Subsequently, linked markers could be employed for indirect selection. Because of the difficulty to predict, which crosses yield inbred lines for the highest performing hybrid, breeders usually employ several crosses and different testers in parallel. Therefore, an important question with respect to MAS is, whether QTL mapping results are consistent across different populations and testers. If not, the costs for MAS strongly increase, because separate QTL mapping is required for each population or tester.

Numerous QTL studies on several traits have already been performed in maize[2]. In most of these investigations selfed or backcrossed progenies were employed. Only few studies compared the QTL mapping results obtained with different testers.[3,4] While Schön et al. (1994)[4] reported fairly consistent results for plant height and kernel weight, Ajmone-Marsan et al. (1994)[3] obtained poor agreement between two testers for grain traits.

In this study, TC progenies of 380 F_3 lines with two testers were employed to map QTL for important forage traits including dry matter yield, dry matter concentration, starch concentration, *in vitro* digestible organic matter, and plant height. The main objective was to investigate the consistency of QTL results across testers.

2 MATERIALS AND METHODS

2.1 Plant Materials

The plant materials used in this study were described in detail previously[4]. Two early maturing elite European flint inbreds (P1, P2) were used as parents to produce 380 randomly chosen $F_{2:3}$ lines derived from the cross P1 x P2. Testcross (TC) seed was produced in two seperate isolation plots using two elite European dent testers (T1, T2) as pollinators.

2.2 RFLP Assays

The procedure for RFLP assays and analyses of marker data have been reported elsewhere[4]. RFLP data and linkage map given by these authors were also employed in this study. The maize genome was covered by 89 well distributed RFLP markers.

2.3 Agronomic Trials

The TC progenies were evaluated in two experiments (Exp. 1 = TC with T1; Exp. 2 = TC with T2). Five agroecological diverse environments (site-year combinations) were employed in each experiment (Exp. 1: Einbeck 1991, Hohenheim 1993, Eckartsweier 1993, Zell 1993, Eckartsweier 1994; Exp. 2: Einbeck 1991, Eckartsweier 1992, Hohenheim 1992, Hohenheim 1993, Eckartsweier 1994). Each experiment included 400 entries: 380 TC of the 380 F_3 lines, TC of P1 and P2 were included as quintuple entries, and 10 checks. The experimental design was a 40-by-10 alpha design with two replications. Plots were overplanted and later thinned to a final plant density of 11 plants m^2. All experiments were machine planted and harvested as forage trials with a chopper.

Data were analysed for the following traits: plant height in cm, measured as distance from soil level to the lowest tassel branch; dry matter concentration (DMC) of forage in g kg^{-1}, determined from a representative sample of 1.5 kg chopped material per plot; dry matter yield (DMY) of forage in Mg ha^{-1}, calculated by multiplying the DMC values with the fresh weight per plot; starch concentration (STC) and *in vitro* digestible organic matter (IVDOM) in g kg^{-1}, measured by near-infrared reflectance spectroscopy (NIRS)[5].

2.4 Data Analysis

Adjusted entry means and effective error mean squares of trait data were used to compute the combined analysis of variance and covariance across environments for each experiment. Variance components and heritabilities were estimated as described elsewhere[4]. The resulting entry means across and for individual environments were employed for QTL mapping using the composite interval mapping (CIM) approach[6,7] implemented in the program PLABQTL[8]. A LOD threshold of 2.5 was chosen for declaring a putative QTL. QTL positions were determined at the LOD maxima in the regions under consideration and a one-LOD support interval was constructed as described by Lander and Botstein (1989)[9]. QTL with non-overlapping support intervals were considered as being different. Putative

QTL were examined for the presence of QTL-by-environment and digenic epistatic interactions[8].

3 RESULTS AND DISCUSSION

Testcross means of F_3 lines across the three environments in common were significantly (P<0.05) greater for T1 than T2 for all traits except DMC. The TC means of P2 significantly (P<0.05) exceeded those of P1 for DMY and STC in both experiments and for DMC in Exp. 1. No difference was observed for IVDOM in both experiments and for DMC in Exp. 2. For all traits, transgression of the parental TC means by the TC means of F_3 lines occurred in both directions. The orthogonal contrast between the average TC performance of the parent lines and the TC mean of the 380 F_3 lines revealed significantly (P<0.05) higher means of the latter for DMY and PHT in both experiments. Therefore, at the level of generation means the presence of epistasis was indicated only for DMY and PHT.

Genotypic variance components were significant (P<0.05) for all traits and variance components for genotype-by-environment interactions were significant for all traits except IVDOM in Exp. 2. Heritabilities ranged between 0.64 and 0.88 for DMY, DMC, PHT, and STC. The heritabilities for IVDOM were 0.45 in Exp. 1 and 0.24 in Exp. 2. Sample jumbling as reason for low heritabilities of IVDOM can be ruled out because of high heritabilities for STC, measured from the same NIRS samples. In conclusion, either P1 and P2 are genetically similar in respect to IVDOM or both testers mask an important part of the difference between both parental lines.

Phenotypic correlations between TC of F_3 lines across the three common environments in Exp. 1 and Exp. 2 ranged from 0.28 to 0.67 and were highly significant (P<0.01) for all traits. Genotypic correlations exceeded 0.80 and corresponding phenotypic correlations in all cases.

Table 1 *Genotypic Correlations in Exp. 2 (Below Diagonal); Above Diagonal: Number of "Common QTL" for Two Traits (First Integer); Number of those "Common QTL", where the Allele, Increasing Trait Expression Originates from the Same Parent (Second Integer).*

Trait	DMY	DMC	PHT	STC	IVDOM
DMY		6 / 2	4 / 4	4 / 0	1 / 1
DMC	-0.02		5 / 1	3 / 3	4 / 3
PHT	0.73**	-0.15**		3 / 0	4 / 1
STC	-0.45**	0.58**	-0.53**		1 / 0
IVDOM	-0.01	0.11	-0.30**	0.09	

** *Genotypic correlation exceeded two times its standard error.*

Table 2 *Number of QTL detected in Exp. 1 (T1) and Exp. 2; Common QTL across Testers (T1 ∩ T2); Frequency of QTL-by-Environment Interactions and Digenic Epistasis*

	Number of QTL			QTL x E		Epistasis	
Trait	T1	T2	T1 ∩ T2	T1	T2	T1	T2
DMY	10	10	3	5	1	0	0
DMC	9	11	8	1	0	0	0
PHT	16	14	10	0	1	0	1
STC	9	10	4	0	4	0	0
IVDOM	7	6	2	1	0	0	1

Genotypic correlations among traits were similar for Exp. 1 and Exp. 2 and reflected at the QTL level. In Exp. 2, close positive correlations (0.73) were detected between DMY and PHT (Table 1). Four genomic regions affected simultaneously DMY and PHT (Table 1). In all four cases, the DMY increasing allele also increased PHT, as expected for positively correlated traits. Likewise, the negatively correlated traits PHT and STC had three genomic regions in common. In all three cases, the PHT increasing allele decreased STC. Although by our approach pleiotropy and close linkage can not be resolved, genotypic correlations can at least partly be explained by common QTL. However, even in the case of closely correlated traits like DMY and PHT, the majority of QTL was detected for only one of both traits (Table 1). In Exp. 2, six regions affected DMY, but not PHT, and 10 chromosome segments were involved in the expression of PHT, but not DMY. Conclusively, within a given population and tester combination, markers can efficiently support the simultaneous improvement of negatively correlated characters.

As expected from the heritability estimates, the highest number of 16 QTL was detected for PHT in Exp. 1 (Table 2) and the lowest number of six QTL for IVDOM in Exp. 2. Consistent QTL results across testers were obtained for DMC: eight QTL were in common across both experiments and only one or three QTL were uniquely detected in Exp. 1 or Exp. 2, respectively (Table 2). A comparable level of consistency was detected for PHT (Table 2), indicating a preponderance of additive gene action for these traits. In conclusion, information about QTL positions for DMC and PHT obtained with one tester should also apply to other testers of the same germplasm pool.

QTL results were inconsistent across testers for DMY, STC, and IVDOM despite close genotypic correlations among TC experiments. One reason for this discrepancy might be that the detected QTL explain only about half of the genetic variation. Less than 50% of the QTL identified in one TC experiment were detected in the other TC experiment. Six and four QTL for DMY and STC, respectively, displayed significant ($P<0.05$) QTL-by-environment interactions, explaining at least part of the differences between both experiments. While digenic epistatic interactions were of minor importance, dominance and chance effects might be other factors affecting the consistency of QTL results across testers[4]. Consequently, for DMY, STC, and IVDOM, a separate QTL mapping of TC progenies is required for each tester employed.

In summary, MAS might be an efficient tool to develop the best line within a given population and for a given tester. However, at least for the most important forage traits DMY, STC, and IVDOM, information about QTL positions gained from one TC experiment seems to be of little value for another TC experiment and very likely for other populations.

References

1. J. Boppenmaier, A.E. Melchinger, E. Brunklaus-Jung, H.H. Geiger, R.G. Herrmann, 1992, *Crop Sci.* **32**, 895.
2. M. Lee, 1995, *Adv. Agron.* **55**, 265.
3. P. Ajmone-Marsan, G. Monfredini, W.F. Ludwig, A.E. Melchinger, P. Franceschini, G. Pagnotto, M. Motto, 1994, *Maydica* **39**, 133.
4. C.C. Schön, A.E. Melchinger, J. Boppenmaier, E. Brunklaus-Jung, R.G. Herrmann, J.F. Seitzer, 1994, *Crop Sci.* **34**, 378.
5. H. Degenhardt, PhD Thesis, University of Halle, 1996.
6. R.C. Jansen, P. Stam, 1994, *Genetics* **136**, 1447.
7. Z.-B. Zeng, 1994, *Genetics* **136**, 1457.
8. H.F. Utz, A.E. Melchinger, 1996, *JQTL* **2**, 1.
9. E.S. Lander, D. Botstein, 1989, *Genetics* **121**, 185.

SESSION III
MAIZE BIOTECHNOLOGY: RECENT DEVELOPMENTS IN MAIZE TRANSFORMATION APPLICATIONS

Maize Biotechnology: Recent Developments in Maize Transformation and Applications

Paul Christou

LABORATORY FOR TRANSGENIC TECHNOLOGY & METABOLIC PATHWAY ENGINEERING,
JOHN INNES CENTRE, NORWICH NR4 7UH, UK

1 INTRODUCTION

Maize has been the model system for fundamental genetic studies for many decades. The immense wealth of information on maize genetics and the global importance of the crop for food, feed and industrial applications, resulted in intense research in terms of classical breeding for the conventional improvement of the crop. Incorporation of biotechnology approaches to the improvement of maize was a natural consequence. Tremendous resources have been devoted to the genetic engineering of maize, both in industry and in academia. This concentration of resources resulted in the creation and commercialization of transgenic maize germplasm expressing a number of different agronomic traits. Insect and herbicide resistances, male sterility for hybrid seed production and modification of metabolic properties are first generation traits. More complex properties, including fungal and bacterial resistances, photosynthetic properties, applications for alternative uses, stress resistances, etc., are viable targets which are accessible through modern molecular biology. It is no surprise, therefore, that maize biotechnology is the most advanced amongst all cereal crops.

Advances in maize biotechnology resulted from concerted progress in three interdependent areas: molecular biology and gene isolation, cell biology and gene transfer. In this report, we will address these three areas in some detail and we will provide an update of current progress and future prospects for maize biotechnology. Diverse technologies for gene transfer such as particle bombardment and Agrobacterium will be discussed and their relative merits will be highlighted. An update of field trials involving transgenic maize is also given.

2 REVIEW OF MAIZE TRANSFORMATION

Early research activity in maize biotechnology, with emphasis on development of suitable gene transfer systems, started in commercial laboratories, primarily in the US and to a lesser extend in Europe (Pioneer HiBred, AgrEvo, Monsanto, DeKalb Genetics, Ciba-Geigy, Zeneca, Sandoz, Plant Genetic Systems and others). Recovery of maize plants from electroporated transformed protoplasts resulted in a major disappointment when all transgenic plants obtained in the early experiments were shown to be sterile[1]. Agrobacterium was shown to be effective in introducing viral DNA into maize, but was incapable of achieving germline transformation in early experiments[2].

Two groups utilized embryogenic maize suspension cultures for the recovery of transgenic, fertile plants[3,4]. A specific cross was utilized by both groups to develop a regenerable embryogenic suspension culture amenable to transformation. Particle bombardment of such a suspension culture, followed by selection using the herbicide bialaphos, resulted in transformed embryogenic calli in independent experiments. Fertile transgenic plants were regenerated and stable inheritance and expression of bar, and functional activity of the enzyme phosphinothricin acetyltransferase, were observed in a number of subsequent generations. The transformation process and the presence of foreign gene(s) did not affect plant vigor or fertility[5]. Transformation of regenerable maize tissues by electroporation of wounded immature zygotic embryos or embryogenic type I callus provides an alternative to particle bombardment[6]. Tissues were wounded either mechanically or enzymatically and were subsequently electroporated with a neomycin phosphotransferase II gene. Transformed embryogenic calli were selected from electroporated tissues on kanamycin-containing medium and fertile transgenic plants were regenerated. Absence of chimerism and Mendelian segregation in R1 progeny showed that transgenic plants were derived from a single cell or a small number of cells. It is still not possible, however, to generate regenerable cultures from most of the elite maize inbreds. Consequently, further refinements and alternative culture systems had to be developed to make the genetic engineering of maize effective and practical. Lowe et al., described a germline transformation system for maize, following manipulation of chimeric shoot meristems[7]. They investigated whether cells in the developing shoot meristem of immature zygotic embryos might provide an alternative, more universal target for production of transformed maize plants. Following DNA delivery, immature embryos developed into chimeric plants with transgenic sectors containing an antibiotic resistance marker and the gusA gene at a high frequency. Because the majority of transgenic sectors was restricted in size, the probability of a transformation event contributing to the germline without further manipulation was estimated to be very low. To enlarge the transgenic sectors and increase the likelihood of germline transmission, the apical meristems of germinated plants were excised and cultured on cytokinin-containing medium with a selective agent. Transformed sectors were visualized by their non-bleached phenotype or by staining for GUS. Hormonally-induced shoot multiplication produced plants with sectors that had a greater chance of contributing to germline. Transmission to progeny was demonstrated both by transgene expression and by Southern analysis. The authors reported that this method was used successfully with genotypes that included a sweet corn hybrid and an elite field corn inbred.

To develop a less genotype-dependent maize-transformation procedure, Wan et al., used 10-month-old Type I callus as target tissue for microprojectile bombardment[8]. Twelve transgenic callus lines were obtained from two of the three anther-culture-derived callus cultures representing different genetic backgrounds. Multiple fertile transgenic plants (R0) were regenerated from each transgenic callus line. Transgenic leaves treated with the herbicide Basta showed no symptoms, indicating that one of the two introduced genes, bar, was functionally expressing. Data from DNA hybridization analysis confirmed that the introduced genes (bar and gusA) were integrated into the plant genome and that all lines derived from independent transformation events. Transmission of the introduced genes and the functional expression of bar in R1 progeny were confirmed. Germination of R1 immature embryos in the presence of bialaphos was used as a screen for functional expression of bar; however, leaf painting of R1 plants proved a more accurate predictor of bar expression in plants. This study suggests that maize Type I callus can be transformed efficiently through microprojectile bombardment and that fertile transgenic plants can be recovered. This system should facilitate the direct introduction of agronomically important genes into commercial genotypes.

In an attempt to develop a facile and variety-independent gene transfer system for maize, Heng et al., studied the in-vitro morphogenetic pattern of corn (Zea mays L.) shoot tips excised from aseptically-grown seedlings, and of explants of axillary shoot buds, immature tassels and ears (staminate and pistillate inflorescences) obtained from greenhouse-grown corn plants[9]. The seedling shoot tips and immature ears first regenerated clumps of multiple shoots within four weeks of culture. Based on the culture procedures described above, Zhong et al. developed a reproducible system for the recovery of fertile transgenic maize plants[10]. The transformation was carried out by bombarding cultured shoot apices with a plasmid carrying the bar gene and the potato proteinase inhibitor II gene (pin2)[11]. Bombarded shoot apices were multiplied in culture and selected with glufosinate ammonium. Co-transformation frequency was 100% for linked genes and 80% for unlinked genes. This system has distinct advantaged over alternative culture systems for transformation. Shoot tips have been propagated in culture for a very long time (more than 4 years) without loss of regeneration potential. In order to assess expression of the bar gene, the herbicide Ignite[R] was sprayed on all recovered plants. All engineered plans were shown to be resistant to the herbicide whereas untransformed control pants exhibited necrosis in two days and died 10 days after application of the herbicide. Using this system, the authors reported recovery of regenerated plantlets from 36 different genotypes.

Ishida et al.,described a transformation system for maize mediated by Agrobacterium tumefaciens[12]. Transformants of the inbred line A188 were recovered following cocultivation of immature embryos with a "super-binary vector". This is a significant advance in the field; however, it will be important to demonstrate transformation of elite maize germplasm using Agrobacterium before the method can be incorporated into routine programs involving biotechnology improvement of maize.

3 A NOVEL SYSTEM FOR MAIZE TRANSFORMATION

It is apparent that in order to put in place an effective gene transfer system not only for maize but also for any other plant, one of the key requirements is to have in hand a very efficient and rapid regeneration system. Such a system should have the capacity to give large numbers of regenerated plants from tissues which are compatible with a given DNA delivery procedure. In this section, we will describe such a system. The system is based on the observation that isolated meristems from germinated maize seedlings have the capacity to develop into proliferative organogenic or embryogenic cultures depending on the composition of the culture medium. The procedure calls for sterilized seeds of maize to be germinated on basal media supplemented with high levels of cytokinin (BAP at 3 mg/l). Seven to ten days after germination the general area of the meristem swells, as shown in figure 1a. A thin horizontal section (1-3 mm in diameter) excised from the basal end of the germinating seedling exposes a disc which contains the induced meristem area (Fig 1b). This tissue is very plastic and can be regenerated to plants effectively through either a somatic embryogenesis pathway (Fig. 2) or through an organogenic pathway involving repetitive induction of exposed de novo meristems (Fig. 3). Gene transfer experiments may be carried out at various developmental stages to recover transgenic plants. Our work is now focusing on optimization of parameters which control effective recovery of transgenic plants from these tissues. Our experiments also aim to identify the optimum selection agent which would allow facile selection of transgenic plants.

4 STRUCTURE OF INTEGRATED DNA INTO MAIZE

Zea mays transformants produced by particle bombardment of embryogenic suspension culture cells of the genotype A188 x B73 and selected on kanamycin or bialaphos were characterized with respect to transgene integration, expression, and inheritance[13]. Selection on bialaphos, mediated by the bar or par genes, was more efficient than selection on kanamycin, mediated by the nptII gene. Most transformants contained multicopy, single locus, transgene insertion events. A transgene expression cassette was more likely to be rearranged if expression of that gene was not selected for during callus

1

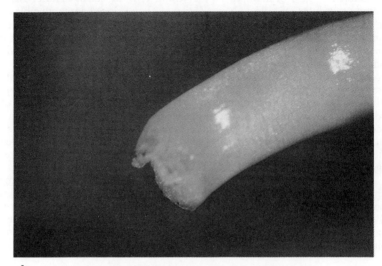

A

B

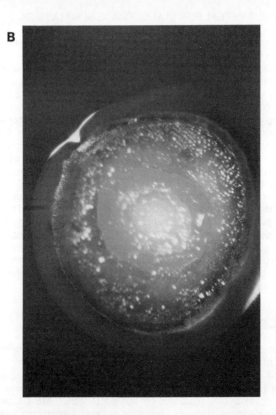

1.   Morphology of maize meristem region following germination of seedlings on media containing high levels of cytokinin.

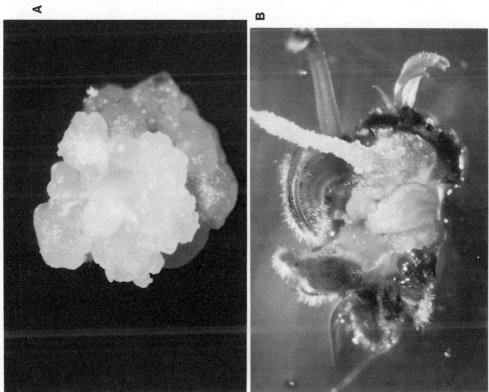

2. Regeneration from maize meristems via a somatic embryogenesis pathway. A: induction of embryogenic callus using 2,4D; B, C: shoot and plantlet formation from embryogenic callus.

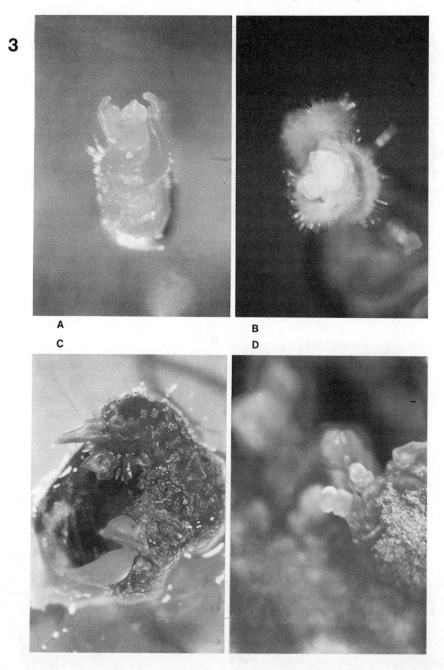

3.   Regeneration from maize meristems using a <u>de novo</u> meristem multiplication system. A: exposed primary meristem excised from seedlings germinated on medium containing high levels of BAP; B: proliferation of multiple meristems from A. C: shoot developmed from <u>de novo</u> meristems; D: Exposed individual <u>de novo</u> meristems.

growth. Not all plants regenerated from calli representing single transformation events expressed the transgenes, and a nonselectable gene (gusA) was expressed in fewer plants than was the selectable transgene. Mendelian inheritance of transgenes consistent with transgene insertion at a single locus was observed for approximately two thirds of the transformants assessed. Transgene expression was typically, but not always, predictable in progeny plants. Transgene silencing, as well as poor transgene transmission to progeny, was observed in some plant lines in which the parent plants had expressed the transgene.

When two separate plasmids containing bar and gusA were used in transformation experiments involving bombardment of embryogenic suspension cultures of maize, co-integration and co-expression frequencies were shown to be 77 and 18% respectively[5]. Progeny from four independently-derived transformation events were analysed extensively to study inheritance and stability of introduced genes[14]. Data from these experiments demonstrated mendelian segregation of functional genes. Functional bar sequences segregated as single units in all four transgenic lines analysed; however, integrated copies of bar appeared to be unstable in one line. Independent segregation of apparently non-functional copies of bar was also observed. GUS-coding sequences, introduced in co-transformation experiments (separate plasmids) were inherited by all four lines and appeared to be linked to functional bar gene(s) in all cases. However, expression of gusA was not detected in any of the four lines.

5 FIELD EVALUATION OF TRANSGENIC MAIZE

The European corn borer (ECB; Ostrinia nubilalis Hubner) is an economically significant pest of corn (Zea mays L). The ability to routinely transform corn has broadened the control options available to include the introduction of resistance genes from sexually incompatible species. Recovery of transgenic maize, expressing a number of genes of agronomic interest, including insect resistance, resulted in successful field trials with transgenic plants showing excellent resistance to BT. Koziel et al., introduced a synthetic gene encoding a truncated version of the CryIA(b) protein derived from Bacillus thuringiensis into immature embryos of an elite line of maize via particle bombardment[15]. High levels of protein were obtained. Hybrid maize plants resulting from crosses of transgenic elite inbred plants with commercial inbred lines were evaluated for resistance to the European corn borer under field conditions. Plants expressing high levels of the insecticidal protein exhibited excellent resistance to repeated heavy infestations of the insect. Microprojectile bombardment was used to introduce synthetic versions of cryIA insecticidal protein genes from Bacillus thuringiensis subsp. kurstaki (Btk) into embryogenic tissue of the Hi-II (A188/B73 derivative) genotype of corn[16]. Of 715 independent transgenic calli produced, 314 (44%) had insecticidal activity against tobacco hornworm (Manduca sexta L.) larvae. Plants were regenerated, self-pollinated when possible, and crossed to B73. First-generation progeny of 173 independent Bt-protein expressing calli were evaluated under field conditions with artificial ECB infestations in 1992 or 1993. Approximately half (89/173) segregated in a single-gene manner for resistance to first- generation ECB leaf-feeding damage. All of the 89 lines evaluated in 1992 or 1993 for resistance to second-generation ECB exhibited less stalk tunnelling damage than the non- transgenic controls. In 1993, 44% (34/77) of the lines tested had less than or equal to 2.5 cm of tunnelling, compared with severe damage (mean = 45.7 cm) in the B73 X Hi-II controls.

A number of additional field trials including herbicide resistance have been carried out successfully, including one by the company AgrEvo in which maize plants engineered for resistance to the herbicide glufosinate were evaluated successfully in the field (personal communication) and a series of trials by Plant Genetic Systems (Gent, Belgium) evaluating engineered male sterility in maize (personal communication).

6 CONSUMER ACCEPTANCE AND REGULATIONS

Products of plant genetic engineering are rapidly approaching commercialization. A number of engineered crops, including the first transgenic cereals are now in the market. The first cereal crop to be commercialized is maize engineered for insect and herbicide tolerance. We expect to see additional traits introduced into maize to reach the market in the next year or two. The US has taken a leadership role in the commercialization of products of genetic engineering and indeed there now exists a climate which is conducive to deregulation of a number of engineered crops. A number of companies have petitioned successfully for the unrestricted use of transgenic crops, including cereals such as maize, engineered with the Bt gene for resistance against the European corn borer (Ciba-Geigy and others) as well as herbicide resistance (glufosinate and glyphosate, AgrEvo and Monsanto, respectively). The climate in Europe, however, is very different. A number of active environmental groups and non-government organizations (NGOs) have been very successful in introducing very restrictive guidelines for the release of transgenic crops. These groups have been very successful in influencing the general public and there are indeed various levels of scepticism and concern amongst different European countries as to the safety and acceptance of products of recombinant DNA technology. Thus, a clear dichotomy exists between consumer acceptance in the US versus Europe. Under pressure from Britain, the European Union has for the first time blocked the approval of genetically engineered maize, designed to resist the European corn borer, which devastates maize all over the world[17]. These plants were also engineered with a herbicide resistance gene, allowing farmers to control weeds when necessary. The plant has already been approved for sale in the US and Canada. Companies that want to market their products in Europe only need the approval of one country. The findings of that country's regulatory authorities are passed on to their counterparts in all other European countries, and then governments vote to decide whether the application should go unchallenged. France approved the transgenic maize plant, but Britain, Sweden, Austria and Denmark voted it down, with four other countries-Germany, Italy, Luxembourg and Greece-abstaining. Decision now rests with the Council of Ministers. It is interesting to note that reasons for rejecting the product in various European countries were very different. Britain objected due to the presence of an antibiotic resistance gene in the transgenic plant, which precipitated fears that this gene might be picked up by bacteria thus making its use as an antibiotic ineffective. This argument is not necessarily correct, and it is not supported by scientific facts. Sweden, Austria and Denmark objected on the grounds that the maize plants should have been labelled as genetically engineered. They also expressed additional concerns, including the fact that insects might develop resistance to the Bt toxin, and that the gene for herbicide resistance might spread to weeds. It is beyond the scope of this manuscript to argue for or against these views. It is interesting, however, to note the mixed response that this decision precipitated among various constituencies in Europe. Organizations such as Greenpeace, saluted the rejection, characterizing it as a step in the right direction. European biotechnology companies were alarmed, arguing strongly that Europe's strict laws, often not based on rational scientific facts, will handicap European industry, while rivals in the US and Japan enjoy laxer rules[17].

7 FUTURE PROSPECTS

Advances in the genetic engineering of maize lead those for other crops. Such targets may be classified in three broad categories: Agronomic characteristics include mostly single gene traits such as control of weeds, resistance to insect pests and diseases caused by bacteria, fungi or viruses. More complex agronomic characteristics which are controlled by multi-gene families include tolerance to biotic stresses such as heat, cold, drought, salt, etc., and also tolerance to abiotic stresses such as heavy metals etc. Hybrid technology will also benefit from wider applications of genetic engineering, particularly the concept of molecular apomixis. Food processing traits include enhanced nutritional quality of food crops, delayed ripening of fruits, modifications in color, flavor, texture, increased solids for processing vegetables such as tomatoes, as well as modification of oil and starch composition. Safety issues can also be addressed by eliminating toxic or anti-nutritional factors from food products. Finally, industrial uses include the creation of modified and specialty oils, alternative uses of crops to include production of industrial and specialty enzymes and proteins, and also the production of recombinant macromolecules such as antibodies and vaccines for human and animal healthcare. These targets need to be selected carefully, however, to assure that the specific macromolecules in question are produced economically and more efficiently in plants as compared to microorganisms, mammalian cultures or transgenic animals.

References

1. C.A. Rhodes, D.A. Pierce, I.J. Mettler, D. Mascarenhas, J.J. Detmer, *Science*, 1988, **240**, 204.
2. N. Grimsley, T. Hohn, J.W. Daview and B. Hohn, *Nature*, 1987, **325**, 177.
3. M. Fromm, F. Morrish, C. Armstrong, R. Williams, J. Thomas, and T. Klein, *Bio/Technology*, 1990, **8**, 833.
4. W.J. Gordon-Kamm, T.M.Spencer, M.L. Mangano, T.R. Adams, R.J. Daines, W.G. Start, J.V. O'Brien, S.A. Chambers, W.R. Adams Jr., N.G. Willetts, T.B. Rice, C.J. Mackey, R.W. Krueger, A.P. Kausch and P.G. Lemaux, *The Plant Cell*, 1990, **2**, 603.
5. W.J. Gordon-Kamm, T.M. Spencer, J.V. O'Brien, W.G. Start, R.J. Daines, T.R. Adams, M.L. Mangano, S.A. Chambers, S.J. Zachwieja, N.G. Willetts, W.R. Adams Jr., C.J. Mackey, R.W. Krueger, A.P. Kausch and P.G. Lemaux, *In Vitro Cell Dev. Biol.*, 1991, **27**, 21.
6. K. D'Halluin, E. Bonne, M. Bossut, M. De Beuckeleer and J. Leemans, *The Plant Cell*, 1992, **4**, 1495.
7. K. Lowe, B. Bowen, G. Hoerster, M. Ross, D. Bond, D. Pierce and B. Gordon-Kamm, *Bio/Technology*, 1995, **13**, 677.
8. Y. Wan, J.M. Widholm and P.G. Lemaux, *Planta*, 1995, **196**, 7.
9. Z. Heng, C. Srinivasan and M.B. Sticklen, *Planta*, 1992, **187**, 483.
10. H. Zhong, B. Sun, D. Warkentin, S. Zhang, R. Wu, T. Wu and M.B. Sticklen, *Plant Physiol.*, 1996, **110**, 1097.
11. M. Keil, J. Sanchez-Serrano, J. Schell and L. Wilmitzer, *Nucl. Acids Res.*, 1986, **14**, 5641.
12. Y. Ishida, H. Saito, S. Ohta, Y. Hiei, T. Komari and T. Kumashiro, *Nature Biotechnology*, 1996, **14**, 745.
13. J. Register III, D. Peterson, P. Bell, W. Bullock, I. Evans, B. Frame, A. Greenland, N. Higgs, I. Jepson, S. Jiao, C. Lewnau, J. Sillick and H. Wilson, *Plant Molecular Biology*, 1994, **25**, 951.
14. T.M. Spencer, J.V. O'Brien, W.G. Start, T.R. Adams, W.J. Gordon-Kamm and P.G. Lemaux, *Plant Molecular Biology*, 1992, **18**, 201.
15. M.G. Koziel, G.L. Beland, C. Bowman, N.B. Carozzi, R. Crenshaw, L. Crossland, J. Dawson, N. Desai, M. Hill, S. Kadwell, K. Launis, K. Lewis, D. Maddox, K. McPherson, M.R. Meghji, E. Merlin, R. Rhodes, G.W. Warren, M. Wright and S.V. Evola, *Bio/Technology*, 1993, **11**, 194.
16. C. Armstrong, G. Parker, J. Pershing, S. Brown, P. Sanders, D. Duncan, T. Stone, D. Dean, D. DeBoer, J. Hart, A. Howe, F. Morrish, M. Pajeau, W. Petersen, B. Reich, R. Rodriguez, C. Santino, S. Sato, W. Schuler, S. Sims, S. Stehling, L. Tarochione and M. Fromm, *Crop Sci.*, 1995, **35**, 550.
17. A. Coghlan, *New Scientist*, 4 May 1996, p.7.

Approaches to the Genetic Transformation of Sorghum

Ana M. Casas,[1] Andrzej K. Kononowicz,[2] Usha B. Zehr,[3] Lanying Zhang,[2] Theresa G. Haan,[2] Dwight T. Tomes,[4] Ray A. Bressan[2] and Paul M. Hasegawa[2]

[1] DEPARTMENTO DE GENÉTICA Y PRODUCCIÓN VEGETAL, ESTACIÓN EXPERIMENTAL DE AULA DEI (CSIC), APARTADO 202, 50080 ZARAGOZA, SPAIN
[2] CENTER FOR PLANT ENVIRONMENTAL STRESS PHYSIOLOGY, 1165 HORTICULTURE BUILDING, PURDUE UNIVERSITY, WEST LAFAYETTE, IN 47907, USA
[3] DEPARTMENT OF AGRONOMY, PURDUE UNIVERSITY, WEST LAFAYETTE, IN 47907, USA
[4] PIONEER HI-BRED INTERNATIONAL INC., JOHNSTON, IA 50131-1004, USA

1 INTRODUCTION

Sorghum, *Sorghum bicolor* L. Moench, ranks fifth worldwide among cereal crops in area and production. Although production in Europe is scarce, with only 1% of the total 60 million tons in 1994[1], this is a major food and livestock feed crop in Africa, Asia, and parts of North and South America.

Genetic improvement of sorghum for agronomic and quality traits, up until recently, has been carried out by traditional plant breeding methods and improved cultural management practices. Currently new biotechnology approaches are being developed that could have a major impact on crop improvement. Among those, genetic transformation gives access to genes, for crop improvement, derived from recombinant DNA technology regardless of phylogenetic distance. During the last few years, advances in tissue culture and transformation technologies have resulted in the production of transgenic plants of all major cereals[2-5], including sorghum[6-7]. A key to this transformation was the development of microprojectile bombardment devices for DNA delivery into cells with high morphogenic potential[8-10]. Most of the successful transformations have relied on embryogenic or highly morphogenic tissues, such as embryogenic cell suspensions, embryogenic calli, scutellum of immature embryos or meristems[2-5].

This paper describes genetic transformation of sorghum using microprojectile bombardment of both immature embryos[6] or immature inflorescences[7].

2 PLANT MATERIAL

A critical aspect of the transformation process is the choice of the tissue or organ to be used for transformation. An essential consideration is that the final goal is to obtain transformed plants, that are phenotypically normal, and are able to transmit the introduced genes to the progeny. Tissue culture studies on sorghum have described regeneration from a wide range of tissues, such as immature embryos[11], immature inflorescences[12,13], leaf segments[14] or shoot tips[15]. In our transformation research, immature embryos and immature inflorescences were utilized as explants for particle bombardment and plant regeneration. Protocols for regeneration via the embryogenesis pathway from both types of explants had been developed in the laboratory of

Dr.L.G.Butler (Dept. of Biochemistry, Purdue University, USA) and those were the basis for our work.

Sorghum genotypes representing a range of genetic backgrounds and a variety of agronomic types were evaluated for their morphogenic competence using those two explants. Substantial differences in morphogenic response were found among genotypes. The most responsive genotypes were selected for further transformation research. Thus, we focused on immature embryos of P898012, a drought tolerant cultivar, and immature inflorescences of SRN39, a Striga-resistant cultivar.

3 METHODOLOGY

The protocol for induction of embryogenesis and initiation of embryogenic calli from immature embryos[6] included culture of immature zygotic embryos (12-15 days after pollination) onto a medium containing Murashige and Skoog mineral nutrients[16]; organic basal supplements of 10 mg/L thyamine.HCl, 1 mg/L pyridoxine.HCl, 1 mg/L nicotinic acid, 4 mg/L glycine, and 100 mg/L myo-inositol; 150 mg/L asparagine; 10% coconut water; 2.5 mg/L of 2,4-dichlorophenoxyacetic acid (2,4-D); 30 g/L sucrose and 8 g/L agar. Regarding immature inflorescences[7], young inflorescences isolated prior to emergence of the flag leaf, were dissected into segments of about 4 mm. A medium for callus induction contained MS salts, organic basal constituents, 2.5 mg/L 2,4-D, 0.5 mg/L kinetin (KIN), 60 g/L sucrose and 8 g/L agar. Two weeks after initiation of culture from immature embryos, or three to four weeks in the case of inflorescences, calli were transferred to a maintenance medium containing MS salts, organic basal supplements, 2.0 mg/L 2,4-D, 0.5 mg/L KIN, 30 g/L of sucrose and 8 g/L of agar. Morphogenically competent calli were visually selected and subcultured at intervals of 2-3 weeks. Cultures on induction and maintenance media were grown in the dark, at 26°.

Shoots were regenerated from morphogenic callus on medium containing MS salts, organic basal supplements, 1.0 mg/L indole-3-acetic acid, 0.5 mg/L KIN, 20 g/L sucrose and 8 g/L agar. To induce root formation, shoots were transferred to medium containing MS salts at one-half of the standard concentration, 0.5 mg/L of α-napthalenacetic acid, 0.5 mg/L of indole-3-butyric acid, 20 g/L sucrose and 8 g/L of agar. Plant regeneration and shoot and root initiation took place under a 16-hr photoperiod (1000-2000 lx from fluorescent, cool white light).

Microprojectile-mediated DNA delivery to primary explants was accomplished with the Bio-Rad Biolistic PDS 1000/He System. There are several features of the particle gun that influence the efficiency of the transformation process. These include the type, size and quantity of microprojectiles used for bombardment, microprojectile velocity, and target distance, amount of DNA per shot and number of bombardments. Coating of gold or tungsten particles with DNA followed published protocols[10]. Optimal conditions for DNA delivery for each of the primary explants are shown in Table 1.

The plasmids used in this research (pPHP620, pPHP687 and pPHP1528) were provided by Pioneer Hi-Bred International. DNA delivery was evaluated by transient expression of reporter genes[6], i.e. the bacterial *uidA* gene[17] encoding for ß-glucuronidase (GUS) in pPHP620; the maize transcriptional activators R[18] and $C1$[19] that control the synthesis of anthocyanins in pPHP687, or the *luc*[20] firefly luciferase gene in pPHP1528. All the genes were under control of an enhanced CaMV 35S promoter[6].

Table 1 *Particle bombardment conditions*

Parameter	Immature embryos	Immature inflorescences
Size	1-2 mm	1.5-3 cm
Preculture	48 hr	1 to 2 weeks
DNA amount	1-2 µg/shot	2 µg/shot
Particles	Gold 1.5-3.0 µm	Tungsten: 1.0 µm
	500 µg/ shot	125 µg/shot
	Tungsten: 1.7 µm	
	125 µg/shot	
Pressure	1100 psi 1x	1100 psi 2x
Target distance	6 cm	6 cm

A reliable selection procedure is a prerequisite for development of an efficient transformation system. Considering that only a small percentage of the transiently expressing cells are likely to be stably transformed, it is important to use a selection protocol that confers a growth advantage to the transformed cells, while maintaining the competence for regeneration. Initially, the antibiotic kanamycin was used but, as it has been described for other cereal species, sorghum calli grew well on high concentrations of this antibiotic, and became non-embryogenic. An alternative selection system, that has been extremely useful for other cereals employs as selective agent the herbicide bialaphos or Basta, and for this reason it was used in our studies on sorghum. The plasmid used for stable transformation (pPHP620) also contained the selectable marker gene *bar*[21], from *Streptomyces hygroscopicus*, encoding for the enzyme phosphinothricin acetyltransferase (PAT) that inactivates the active compound of this herbicide by acetylation.

As mentioned previously, immature embryos and immature inflorescences were used for transformation. Plants were regenerated from both explants through *in vitro* induced somatic embryogenesis. For each type of explant, it is critical to develop an efficient selection strategy to identify and multiply the transformed cells, without interfering with the developmental process. The selection strategy is a compromise that involves the level of explant organization, the concentration of the selective agent used, and the duration of the selection pressure application. Using the herbicide bialaphos, different selection conditions were applied for each of the primary explants. For immature embryos[6], selection started immediately after bombardment, with a low concentration of herbicide (1 mg/L), on induction medium. Two weeks later, the calli that had just initiated were transferred to maintenance medium containing 3 mg/L. A minimum of three more transfers, two weeks each, on the same concentration of herbicide appeared to be necessary to identify putative transgenic callus. In the case of immature inflorescences[7], even a low concentration of bialaphos inhibited the induction of callus. The material was kept for two weeks on the induction medium without any selection. Then, it was transferred to 3 mg/L bialaphos, subcultured for two weeks while still on induction medium and then to maintenance medium with the same concentration of herbicide. Calli were subcultured on this medium for at least three months while selecting for surviving and embryogenic cells.

Table 2 *Molecular and phenotypic analysis of transgenic plants*

			Insertions			Gene expression		
Plant	Fertility	Phenotype	bar	uidA	luc	PAT	GUS	LUC
1119	fertile	normal	1	5	n.a.	yes	no	n.a.
1409	sterile	abnormal	2	2	n.a.	yes	no	n.a.
1702	fertile	normal	1	1	n.a.	yes	no	n.a.
1752	ms	normal	2	2	n.a.	yes	no	n.a.
2251	fertile	normal	2	n.a.	1	yes	n.a.	no
2441	fertile	normal	3	n.a.	3	yes	n.a.	yes

ms - male sterile; n.a.–not attempted

Calli that survived recurrent selection on the maintenance medium were subsequently transferred to a regeneration medium. Bialaphos was also included in regeneration (3 mg/L) and rooting (1 mg/L) media. Shoots were regenerated from embryo derived material on 3 mg/L bialaphos and several transgenic plants were produced (Table 2). For immature inflorescences, it was necessary to reduce the selection pressure in order to regenerate plants. It was observed that, in the presence of the herbicide, plant regeneration was principally via organogenesis through the differentiation of shoot meristems[7]. Although a really small percentage of calli survived selection during the maintenance stage of culture, numerous plants were obtained on media with 1 mg/L bialaphos, but most of them were not transformed. All the plants that survived Basta (0.6 %) treatment were confirmed to be transgenic by other criteria (Table 2).

In summary, we obtained transgenic plants from 2 sorghum genotypes: 1) P898012, from 1740 immature embryos, 56 plants were regenerated, corresponding to 4 different transformation events (1119, 1409, 2251 and 2441); 2) SRN39, out of 60 bombarded plates of immature inflorescences segments, 191 plants were regenerated, 5 of which were transgenic, corresponding to 2 events (1702 and 1752). The whole process from plating in tissue culture to acclimation in the greenhouse took from 4 to 8 months, depending on the explant.

4 ANALYSIS OF TRANSGENIC PLANTS

Five of six different transformation events produced normal plants, similar to seed-derived sorghum plants (Table 2). These plants were then characterized by molecular (Southern analysis) and biochemical (PAT, GUS and luciferase activity assay) methods. Southern analysis indicated the presence of the introduced genes in all plants that survived Basta treatment, with different patterns of integration. Regarding expression of the genes, all the transgenic plants were positive for herbicide resistance, both *in vivo* as well as *in vitro*, testing for acetylation of PPT[22]. However, we did not observe ß-glucuronidase expression in any of the transformed plants that were studied[6,7]. Since the *uidA* gene was detected by Southern analysis, this result implicates the likelihood of transgene inactivation[23], and that it could be due to methylation. The last two transformations were done using a different reporter gene, *luc* encoding for luciferase and some of the transgenic plants expressed the introduced gene.

Table 3 *Molecular analysis of transgenic plants and herbicide resistance of progeny*

Plant	T_0 plants bar Insertions	T_1 progeny Resistant/Susceptible	Segregation
1119	1	432:117	3:1
1702	1	277:1	256:1
1752	2	78:29	3:1
2251	2	81:27	3:1
2441	3	70:26	3:1

The last step of the transformation process is to confirm the transmission of the introduced genes to the progeny. For the five normal transformation events, the plants were selfed, or backcrossed to normal seed derived control plants. The seeds were collected and planted on soil. When the seedlings were two weeks old, they were sprayed with a diluted solution of herbicide and resistant and susceptible seedlings were scored. In all the cases the herbicide resistance data agreed with Mendelian segregation for 1 or 2 independent loci, with the exception of the progeny from 1702 (Table 3). Transmission of the *bar* gene to T_2 and T_3 generations, and expression in progenies of those five transgenic events, was also confirmed.

We have shown that it is possible to obtain transgenic sorghum plants by microprojectile bombardment of both immature embryos or immature inflorescences. However, these results suggest that it is still too early for application of this technology, in its present state, for routine use in a crop improvement program, due to the low efficiency and genotype dependence of the system. There are several constraints that can be considered to improve the transformation system. Regarding the delivery part, the use of stronger promoters such as the actin1[24] from rice or the ubiquitin1[25] from maize, improve the distribution of particles using baffles or meshes, or use osmotic pretreatments[26]. A very important aspect of the transformation is the tissue culture itself, manipulations in the conditions of tissue culture should improve regeneration, in particular for immature inflorescences as well as alternative selection systems[2-5]. The lack of expression of the reporter *uidA* gene implicates the possibility of transgene inactivation[23] (as a consequence of DNA methylation, rearrangements, genome integration site, or homology dependent gene inactivation), and needs to be carefully evaluated.

Although herbicide resistance was introduced in these transgenic plants, it cannot be considered as a system for routine transformation of sorghum. It is important to keep in mind the possible environmental release of the transgenes due to gene flow from sorghum to wild species[27], such as johnsongrass, *S.halepense*, which would clearly limit the future use of this gene. Selection strategy is one of the priorities of the current research, which is focused on looking for alternative selectable markers based on antibiotics.

ACKNOWLEDGEMENTS

Sorghum research was supported by Pioneer Hi-Bred International (Johnston, IA), and grants from the Consortium for Plant Biotechnology Research Inc., the McKnight Foundation and USAID Grant DAN 254-G-00-002-00 through the International Sorghum and Millet Collaborative Research Support Program.

References

1.   FAO Production Yearbook, Rome, 1994, Vol. 48.
2.   A. M. Casas, A. K. Kononowicz, R. A. Bressan and P. M. Hasegawa, 'Plant Breeding Reviews', (ed) J.Janick, Wiley and Sons, New York, 1995, Vol. 13, Chapter 7, p. 235.
3.   I. K. Vasil, *Plant Mol. Biol.*, 1994, **25**, 925.
4.   A. Jähne, D. Becker and H. Lörz, *Euphytica*, 1995, **85**, 35.
5.   J. Fütterer and I. Potrykus, *Agronomie*, 1995, **15**, 309.
6.   A. M. Casas, A. K. Kononowicz, U. B. Zehr, D. T. Tomes, J. D. Axtell, L. G. Butler, R. A. Bressan and P. M. Hasegawa, *Proc. Natl. Acad. Sci., USA*, 1993, **90**, 11212.
7.   A. M. Casas, A. K. Kononowicz, T. G. Haan, L. Zhang, D. T. Tomes, R. A. Bressan and P. M. Hasegawa, *In Vitro Cell. Dev. Biol. - Plant*, 1997, (in press).
8.   T. M. Klein, E. D. Wolf, R. Wu and J. C. Sanford, *Nature*, 1987, **327**, 70.
9.   J. C. Sanford, *Trends Biotechnol.*, 1988, **6**, 299.
10. J. C. Sanford, F. D. Smith and J. A. Russell, *Methods Enzymol.*, 1993, **217**, 483.
11. D. I. Dunstan, K. C. Short and E. Thomas, *Protoplasma*, 1978, **97**, 251.
12. T. Cai and L. G. Butler, *Plant Cell Tiss. Org. Cult.*, 1990, **20**, 101.
13. R. I. S. Brettell, W. Wernicke and E. Thomas, *Protoplasma*, 1980, **104**, 141.
14. W. Wernicke, I. Potrykus and E. Thomas, *Protoplasma*, 1982, **111**, 53.
15. S. Bhaskaran and R. H. Smith, *In Vitro Cell. Dev. Biol.*, 1988, **24**, 65.
16. T. Murashige and F. Skoog, *Physiol. Plant.*, 1962, **15**, 473.
17. R. A. Jefferson, T. A. Kavanagh and M. W. Bevan, *EMBO J.*, 1987, **6**, 3901.
18. S. R. Ludwig, K. F. Habera, S. L. Dellaporta and S. R. Wessler, *Proc. Natl. Acad. Sci., USA*, 1989, **86**, 7092.
19. J. Paz-Ares, D. Ghosal, U. Wienand, P. A. Peterson and H. Saedler, *EMBO J.*, 1987, **6**, 3553.
20. D. Ow, K. V. Wood, M. DeLuca, J. R. DeWet, D. R. Helinski and S. H. Howell, *Science*, 1986, **234**, 856.
21. C. J. Thompson, N. R. Moyva, R. Tizard, R. Crameri, J. E. Davies, M. Lauwereys and J. Botterman, *EMBO J.*, 1987, **6**, 2519.
22. M. E. DeBlock, J. Botterman, M. Vanderwiele, J. Dockx, C. Thoen, V. Gossele, N. R. Moyva, C. Thompson, M. Van Montagu and J. Leemans, *EMBO J.*, 1987, **6**, 2513.
23. J. Finnegan and D. McElroy, *Bio/Technol.*, 1994, **12**, 883.
24. D. McElroy, W. Zhang, J. Cao and R. Wu, *Plant Cell*, 1990, **2**, 163.
25. A. J. Christensen, R. A. Sharrock and P. H. Quail, *Plant Mol. Biol.*, 1992, **18**, 675.
26. P. Vain, M. D. McMullen and J. J. Finer, *Plant Cell Rep.*, 1993, **12**, 84.
27. P. E. Arriola and N. C. Ellstrand, *Amer. J. Bot.*, 1996, **83**, 1153.

Transformation of the Maize Apical Meristem: Transgenic Sector Reorganization and Germline Transmission

K. Lowe, M. Ross G. Sandahl, M. Miller, G. Hoerster, L. Church, L. Tagliani, D. Bond and W. Gordon-Kamm

PIONEER HI-BRED INTERNATIONAL INC., JOHNSTON, IA 50131-1004, USA

1 ABSTRACT

After particle-mediated DNA delivery into the maize meristem, two approaches were used to increase the likelihood of transgene inheritance. The first method involved extensive meristem reorganization during the selection regime. This approach entailed bombardment of coleoptilar stage embryos with a selectable marker (UBI::NPTII::pinII or UBI::aadA::pinII) and the gene encoding GUS, followed by hormonally-induced shoot proliferation and concommitant selective pressure. This caused sufficient reorganization in the meristem, producing stable periclinal or homogeneously transformed plants. One event in an inbred and two separate events in a sweetcorn hybrid have resulted in transgene inheritance. Transmission of the introduced foreign genes and their expression were followed in the inbred through multiple generations. The second approach involved direct germination; after bombardment of meristems in transitional or coleoptilar stage embryos, various treatments to reorganize the meristem and alter transgenic sector fate were tested during maturation and/or germination. Transgenic sectors were monitored by histochemical staining for GUS activity (expression of UBI::uidA::pinII). Without additional treatment, targeting the meristem at early embryonic stages produced frequent transgenic sectors in plants that typically did not persist or transmit to progeny. Treatments used to reorganize the meristem included physical manipulation (i.e. meristem disruption) and application of selective pressure (e.g. bialaphos selection after co-transformation with a UBI::PAT::pinII construct, or streptomycin selection after introduction of a UBI::aadA::pinII construct). Both physical manipulation and selective pressure resulted in larger, more persistent transgenic sectors. Using this method, transgene expression has been observed in the tassel and in pollen, and recently two separate events containing the aadA gene and GUS have transmitted to progeny.

2 BACKGROUND

The potential value of transforming the apical meristem has been recognized for many years in the plant biotechnology arena, and as with other crops, attempts have been made to transform the maize apical meristem (Cao et al., 1990, Gould et al. 1991). Many of these studies started by targeting the meristem in imbibed or germinating seeds, which contained mature embryos with well-developed shoot apicies covered by sheathing leaves.

The chimeric sectors produced under these conditions would be unlikely (as predicted by fate mapping) to pass the transgenes to progeny (Cao et al., 1990).

One potential solution is to use earlier-staged embryos in which the exposed meristem is comprised of fewer cells. Accordingly, transgenic sectors were produced at high frequencies (i.e. 4 - 45%) for maize after particle-mediated introduction of marker genes into the apical meristem of coleoptilar-staged zygotic embryos (see Figure 1 and Lowe et al., 1995). Sectored plants have been produced in all maize inbred lines tested, suggesting that this aspect of meristem transformation will work across all maize germplasm. Although reducing the size of targeted meristems increases sector frequency and size in the chimeric T0 plant, this alone was insufficient to produce transgene inheritance. Thus, sector size must be even larger to increase the probability of transmission to progeny.

Multiple shoot proliferation on high cytokinin-containing medium was tested as one potential meristem reorganizing treatment. Meristems of coleoptilar-staged embryos were bombarded with marker genes (either NPT-II + GUS, or *aad*A + GUS) and the embryos were germinated. After initial germination, the small shoots were moved onto a multiple shoot-inducing medium including the appropriate selective antibiotic (kanamycin or streptomycin, respectively). Using this method, transgenic shoots have been recovered from a sweetcorn hybrid (Honey N Pearl) and a proprietary Pioneer inbred (Lowe et al., 1995). For both examples, Mendelian inheritance and stable transgene expression through multiple generations have been observed.

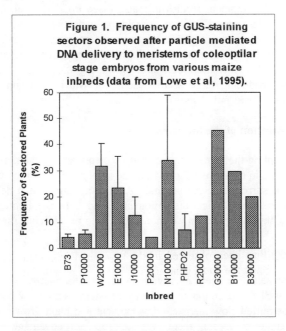

Figure 1. Frequency of GUS-staining sectors observed after particle mediated DNA delivery to meristems of coleoptilar stage embryos from various maize inbreds (data from Lowe et al, 1995).

These experiments demonstrate that cells in the apical meristem of maize are capable of integrating introduced DNA, and that progeny expressing transgenes can be obtained from chimeric shoot meristems with appropriate manipulation. In this case, the manipulation that resulted in reorganizing the sector was hormonally-induced shoot multiplication. Unfortunately, multiple shoot proliferation can be laborious with many inbreds (and not all inbreds respond similarly to exogenously applied hormones), so alternative routes were explored.

3 RECENT PROGRESS

One of the salient features of meristem bombardment is the production of chimeras. Researchers working with crops such as soybean, (Christou and McCabe, 1992), cotton (McCabe and Martinell, 1993) and sunflower (Bidney et al., 1992) have been able to exploit the branching growth pattern and predictable nature of transgenic sectors in these species. Our research to recover transgenic progeny suggested that some degree of transgenic sector reorganization was essential for maize, which required simple, more rapid methods of enlarging sectors in the meristem. Various treatments have been tested with the objectives of i) enlarging transgenic sectors, ii) increasing their persistence and iii) increasing inheritance of transgene expression. These include the use of earlier-stage embryos as targets, physical disruption of the meristem, and application of non-lethal selection regimes that are compatible with continued meristem development. These are briefly described below.

3.1 Targeting Earlier-staged Embryos

Based on fate development studies in maize (see Poethig et al., 1990), targeting developmentally earlier staged embryos would be expected to increase the size of transgenic sectors. Our results have been consistent with this prediction. When transitional stage embryos were bombarded, the size of transgenic sectors and their persistence up the plant increased (compared with using coleoptilar-stage embryos as a target). Despite producing large sectors, some of which went up into the tassel, inheritance to progeny was not observed.

3.2 Physical Disruption of the Meristem

Physically disrupting the apical dome of maize coleoptilar-stage embryos using a glass micropipette resulted in cessation of growth in the apex *per se* and the rapid formation of 2-6 new meristems surrounding the original. After particle-mediated delivery of marker genes into the meristem, physical disruption increased both the frequency and size of sectors; but alone, this manipulation was not sufficient to result in inheritance.

3.3 Meristem-compatible Selective Pressure

Bialaphos, the selective agent most commonly used for maize callus transformation has not worked well for selection of chimeric meristem sectors. When concentrations were used that permitted slow meristem development and plant growth (0.5 to 1.0 mg/l for coleoptilar stage embryos), there was no apparent advantage imparted to the transgenic sector (i.e. sector size did not increase). As bialaphos concentrations were increased, sectored plants that rooted were recovered but none to date have passed transgenes to progeny.

In contrast, a non-lethal selective agent such as streptomycin at 100 mg/l for 2-4 weeks has resulted in plants with large, persistent streptomycin-resistant, GUS+ sectors. Using streptomycin has also permitted unambiguous identification of transgenic sectors (i.e. dark-green leaf phenotype on an albino background) from which plants can readily recover. Recent experiments have produced sectors in T0 plants that persist into the tassel, and preliminary results in a Pioneer proprietary inbred, J10000, and in Hi-II hybrid germplasm have demonstrated both transmission of transgenes and expression of these genes in progeny.

4 CONCLUSION

Our data suggest that milder forms of meristem reorganization (i.e. meristem-compatible selective pressure as opposed to more radical hormonally-induced shoot multiplication) may be sufficient to shift the transgenic sector into a lineage that will contribute to the reproductive tissues. Why is this important, when callus-based transformation is so routine? For many species and/or varieties, *in vitro* culture methods remain onerous, laborious or even impossible. Targeting the meristem and proceeding through a normal developmental pathway should greatly increase the range of germplasm that can be transformed.

It is also important to put this into perspective relative to callus-based maize transformation. Transformation methods for immature embryos and embryo-derived callus required many years of research effort, partially just to learn and understand the biology of the system and how it could be manipulated. Similar to 10 years past with callus-based transformation, perhaps we are at the threshold of learning how to efficiently transform maize meristems. This will require both method development and careful observation to establish the predictability necessary to make this efficient. Streptomycin selection (and other selective agents) in conjunction with some forms of physical meristem disruption that have been demonstrated to increase sector size, will hopefully lead to further improvements.

References

1. D.L. Bidney, C.J. Schelonge and J.B. Malone-Shoneberg, "Proceedings of the 13th International Sunflower Conference", International Sunflower Association, Pisa, 1992, p 1408.
2. J. Cao, Y.C. Wang, T.M. Klein, J. Sanford and R. Wu, "Plant Gene Transfer", C.J. Lamb and R.J Beachy, eds., Wiley-Liss, N.Y., 1990, p 21.
3. P. Christou and D.E. McCabe, *The Plant Journal*, 1992, **2(3)**, 283.
4. J. Gould, M. Devey, O. Hasegawa, E.C. Ulian, G. Peterson and R. Smith, *Plant Physiol*, 1991, **95**, 426.
5. K. Lowe, B. Bowen, G. Hoerster, M. Ross, D. Bond, D.Pierce and B. Gordon-Kamm, *Bio/Technol.*, 1995, **13**, 677.
6. R.S. Poethig, C.N. McDaniel and E.H. Coe, Jr., "Genetics of Pattern Formation and Growth Control", A. Mahowald, ed., Wiley-Liss, N.Y., 1990, p 197.
7. D.E. McCabe and B.J. Martinell, *Bio/Technol.*, 1993, **11**, 596.

Maize as Host for *Agrobacterium*

K. Konstantinov, S. Mladenović-Drinić, T.Ćorić, D. Ignjatović,
G. Saratlić and J. Dumanović

MAIZE RESEARCH INSTITUTE, SLOBODANA BAJIĆA 1, ZEMUN POLJE, 11081 ZEMUN-
BELGRADE, YUGOSLAVIA

1 INTRODUCTION

To change the genetic properties of crops and optimise them for agricultural purposes breeders are trying to combine genomes from different but related cultivars, varieties or species.At the very early period of investigation how recombinant DNA technology could be applied in plant breeding, among several steps, the important one was development of the proper vectors for foreign gene introduction into the plant cell [1]. First, genes dealt with by each scientific group independent of the method used for plant cell transformation, have been appropriate selectable markers for transformants identification [2]. Potrycus [3] reviewed almost all possible approaches for gene transfer into higher plants. Results of different experimental procedures successful in maize transformation are mainly based on the foreign genes introduction by the genetic manipulations of cell or tissue culture [4,5,6,7,8,9]. Results obtained in our first experiments on maize transformation, started during the summer 1986 using *neomycinphosphotransferase* (NPT II) coding sequence as selectable marker have been reported, 1988 [10]. Next, experiments were started during the summer of 1988 using NPT II gene, *A.tumefaciens* wild type strain B6S3 and *cat* coding sequence (each of them provided by J.Schell , R.Topfer and M.Frey, Max-Planck-Institute für Pflanzenzuchtung, Cologne, Germany). Results obtained after analysis of T4, T5 and T6 progeny of transgenic plants enable us to conclude that foreign DNAs have been both integrated, transmitted through the cell cycles and expressed in the maize transgenic plants. In this report, results on bacterial gene effect on embryo protein complexes of transgenic inbred lines are discussed. Also, results obtained in the field trials set up to test combining ability of transgenic inbreds will be presented.

2. EXPERIMENTAL PROCEDURES

2.1.1. Bacterial and plant DNA preparation. Transformation of competent *E.coli* cells by plasmids carrying NPT II and *cat* gene, growth of bacterial culture and purification of plasmid DNA and coding sequences were performed according to standard procedures[11]. *A.tumefaciens* bacterial B6S3 strain has been grown and both chromosomal and plasmid DNA purified according to De Block et al.[12]. Plant DNA was isolated according to Wienand et all.[13] and hybridised to labelled probe according to Southern [14].

    2.1.2. Pollen tube pathway transformation procedure. Essentially the same approach as Ohta published[15] was followed. Pollen viability was tested, by analysing "pollen popping" in a buffer solution using standard light microscopy procedures. It was found that 30 minutes is a critical period for 50% pollen grains to stay unpopped before drifting on to the silk. Pollen grains were pasted in 1.5 ml Ependorf tubes in the buffer for pollen germination as follows: $H_3 BO_3$, 150mg/l; $Ca(NO_3)_2$, 300mg/l; $MgSO_4$, 200mg/l; KNO_3, 100mg/l and sucrose, 150mg/l. The buffered plasmid DNA / pollen grain ratio (w/w) was always 10 micrograms DNA/mg of pollen grains and microliters of DNA solution/mg of pollen grains (v/w) was 10 microliters of DNA solution per 1 mg of pollen grains. Plants, selfpollinated by their own "pasted" pollen grains were grown till maturation and labelled as T_0 plants.

    2.1.3 Incubation of dry kernels in pRT100 plasmid DNA solution. a) Germination. Surface sterilised maize kernels of inbred line B73 were incubated in plasmid DNA solution (10 micrograms per kernel) for 30 minutes at room temp. Incubated kernels were planted on MS medium[16] without hormones, containing plasmid DNA at a concentration of 10 micrograms per ml. Kernels were allowed to germinate and grow for 5 days on this medium in the dark at $25\,^{\circ}C$.

    2.1.4. Transformant selection. Five day old etiolated seedlings were transferred on liquid nutritive solution containing micro and macro elements[16]. Kanamycin (for NPT II) and chloramphenicol (for *Cat* transformants) concentrations in nutritive, liquid medium were 50 mg/l during the first week and increased to 100 mg/l throughout the next four weeks[17]. Green and etiolated leaf tissue enzymatic activity *in situ* assay was performed as for neomycinphosphotransferase[17] and chloramphenicol acetyl transferase[18,19]. Green, NPT II and CAT positive T_0 plants derived from incubated kernels, and T_1 plants derived from the seeds of T_0 pollen tube pathway plants were transferred to field conditions and grown till maturation.

    2.1.5. Allozyme extraction and electrophoresis. Electrophoretic analysis of genotypes at 16 enzyme loci were conducted on crude extracts of coleoptile tissue from 5 day old etiolated seedlings, following the previously standardised procedure[20].

    2.1.6. Field trial testing of the combining ability. Transformants flowering earlier than original inbred B73 were crossed to standard maize tester lines known to have good general and specific combining ability. Data on heterotic effect have been collected on the basis of field trials set up in three replications and three locations in a randomised block design. As control for "new" hybrids, elite single crosses of different maturation rates were included in the field trials.

    2.1.7 Protein extraction and gel electrophoresis. Salt and total soluble proteins were extracted from dry embryos isolated from pre-soaked dry seeds in distilled water for at least four hours at room temperature. Bulk samples, consisting of 10 embryos were analysed. Total soluble proteins have been extracted according to Zivy et al.[21] and salt soluble after Wang et al.[22]. Polyacrylamide gel electrophoresis was performed according to Leammly[23].

3 RESULTS AND DISCUSSION

The idea to improve crop plants by integration of foreign genes, originating either from different plant species or from the organisms belonging to evolutionarily different categories, is as old as recombinant DNA technology used for gene isolation, manipulation *in vitro* and transfer to the plant cell. The transgenic cultivars such as potato, tomato, cotton and recently maize are already evaluated in field trials[24]. All who have evaluated transgenic plants have recognised that transgenes do not always behave only in a predictable way [25,26]. Our results on coding sequence integration and presence of the reporter enzyme, neomycin phosphotransferase (NPT II) in active form in transgenic maize (obtained after pRT100 *neo* plasmid DNA introduction by microinjection into archesporial tissue several cell cycles before meiotic division[10]) showed that it is possible to transform maize genome by direct DNA injection into differentiated tissue[27]. Results obtained in a second set of transformation experiments with NPT II gene are already published in detail [28,29,30].

Results on maize transformation by *A.tumefaciens*, reported previously have been obtained either by injection[31] or agroinfection[32]. Strain–specific lysopine dehydrogenase activity has been demonstrated in B6 infected material [32]. As in our experiments with NPT II gene transfer to maize genome, visible alteration of maize genome expression was identified; the hypothesis was that the same effect for both *A.tumefaciens* and *cat* DNA integration could be expected. This was the reason that during summer of 1988, new experiments with NPT II gene, parallel to plasmid DNA from wild *A.tumefaciens* B6S3 strain and plasmid containing *cat* DNA were started. As expected, *A.tumefaciens* DNA and *cat* DNA integration induced phenotype alteration in T_1 progeny derived from the T_0 plants. Results on efficiency of different transformation technologies in transgenic maize development after NPT II and *A. tumefaciens* DNA are presented in Table 1.

Different phenotypic changes, including barren stalk, rolled leaves, dwarfs, anthocyanin synthesis, kernel texture, zebra striped leaves, stalk height, more vigorous plants, flowering time shifting etc.have been induced both after NPT II, *A.tumefaciens* DNA and *cat* gene integration. Ohta[15] reported the same results after maize transformation by pollen tube pathway. Gordon-Kamm et al.[39] reported that most of R_1 transgenic maize plants regenerated from the calli transformed with bacterial gene *bar*, were more vigorous than the R_0 plants. Many of these altered phenotypes were eliminated through first meiotic division, but effect of *A.tumefaciens* DNA application on flowering time and plant height alteration was noticed in T_1 and T_2 transgenic generation. Data on frequency of transformed plants, which flowered earlier than control inbred line are in agreement with previously obtained results by pollen tube pathway or dry seed incubation in plasmid pRT100*neo*[29].

Plants flowering earlier than control genotype, obtained by each applied technology and three bacterial DNAs, were selected and propagated by selfpollination till 1996, when allozyme profiles and embryo protein complexes were analysed. Allozymes were analysed throughout all plant generations and used as genetic markers to select only genotypes with a characteristic "transgenic" pattern. Results for 1996 generation are presented in Table 2. Obtained results are showing that both *E.coli* and *A. tumefaciens* DNA induced the alteration of the same loci controlling allozyme activity. Stability of

Table 1. *Maize transformation efficiency by different bacterial gene integration*

Genes	DNA injection NPT II	Pollen tube pathway NPT II	B6S3	Dry seed incubation NPT II
% of efficiency	0.1	0.15	0.09	0.13
Transgenic generation	9	8	8	8

Table 2. *Bacterial DNA effect on allozyme genotype alteration in stable transgenic plants*

Genotype	Pgma	Pgmb	Acp1	Mdh1	Mdh2	Phi	Idh2
			Alleles with genotype alteration				
Control -B73	9/9	4/4	2/2	6/6	3.5/3.5	4/4	4/4
B6S3	8/8	4/4					
NPT II		4/4		3.5/6		4/6	
NPT II		4/4	1/6	3.5/6		4/6	
Control - Ms71A	9/9	4/4	4/4	1/6	3.5/3.5	4/4	4/4
Ms71AxB6S3	9/16		2/2	6/6			
Ms71AxB6S3-soft kernel			2/4	6/6			
Ms71AxB6S3- "			2/2			5/5	
Ms71AxB6S3- "							6/6

allozyme profiles throughout plant generations could be taken as reliable indicator of stable foreign gene integration into maize genome. Our results on the relationship between earlier flowering and allozyme pattern are in agreement with those reported by Guse et al.[33] who tried to manipulate allozyme frequencies at the loci conditioning early season vigour and maturity. Also, Dobley et al.[34] have documented allozyme association with early season performance for maize grown at higher attitude.

Transgenic inbred V/460   V / 460 x Mo17

Transgenic V/460 mbred line developed by NPTII gene integration has been crossed to Mo17 inbred. Typical plants are shown in Fig. 1. Small figures are given as example of gene probe lighting of genomic DNAs isolated from leaf tissue of field grown plants. Results obtained in the field trials after crossing of earlier transgenic lines to standard maize

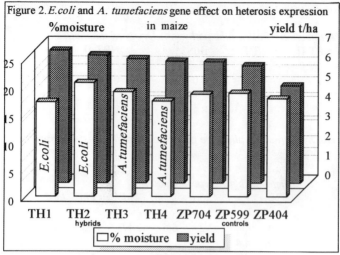

Figure 2. *E.coli* and *A. tumefaciens* gene effect on heterosis expression in maize

inbreds for testing of combining ability are presented in Fig. 2. Standard crosses of high vigor ZP404, ZP599 and ZP704, belonging to different FAO groups were used as controls. Undoubtedly, earlier flowering of transgenic lines has hereditary character and is expressed in hybrids of their crosses to standard tester lines.

If general objectives of new methods are to identify genes or DNA sequences that affect plant development and productivity, our results offer model plants for identification and localization of loci controlling duration of developmental phase of maize pollination and its relation to loci controlling allozyme activity. Loci coding malate dehydrogenase and acid phosphatase are suggested to be candidates for such studies because it looks likely that they are good targets for foreign DNA visiting or integration. Longo and Scandalios[35] reported that both cytosolic and mitochondrial forms of *Mdh* are active as dimmers.   They also suggested dimer formation between subunits localized on the same but not between different subcellular compartments. Probably, changes of *Mdh* activity we observed in transgenic plants could be the consequence of altering both the cytosolic and mitochondrial forms. Abler and Scandalios[36] suggested the same mechanism for induced *Cat*-2 null mutation. They found that this mutation is the consequence of DNA sequence insertion into *Cat*-2 gene. The same type of event occurred by MuI transposon insertion into the first intron of a maize *Adh* gene[37]. As both enzyme forms are encoded by nuclear loci and translated on cytoplasmatic polyribosomes[38] our results indicated that insertion of both *E.coli* and *A.tumefaciens* DNAs occurred at several independent sites in the maize genome. Perhaps this could explain appearance of very broad spectra or phenotype alteration in transgenic plants [15, 29].

Further experiments were focused to determine whether transgenic genotype could be distinguished from the original by determination of gene expression products in dry embryo tissue. Total soluble proteins have been chosen as possible biochemical markers[15,40]. Electrophoregrams of total soluble proteins, isolated from transgenic lines transformed by B6S3 *A.tumefaciens* DNA and *cat* gene, are presented in Plate I.

Results clearly support the idea that bacterial gene integration has an effect on host genome expression by the induction of new polypeptide synthesis (arrows). Results on the effect of NPT II, *A.tumefaciens* (B6S3) and *cat* DNA on embryo soluble protein alteration are presented in Plate II.

Particular polypeptides indicated by arrows are suggested as genetic markers for transgenic genotype identification either as inbred lines or components in hybrid combination. These new proteins are also promising tools for theoretical investigation of genetic control of foreign gene integration and expression both in maize and other monocotyledonous plants.

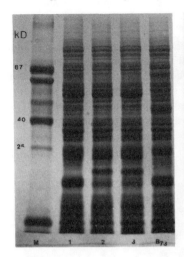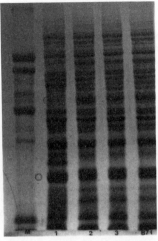

Plate I. A. *A.tumefaciens* DNA effect on soluble protein alteration in maize dry embryo: line M, m.w.marker; lines 1,2 and 3 transgenic B73 inbred; B73 as control.
B. *cat* gene effect on soluble protein alteration in maize dry embryo: M,m.w.marker; line 1 EMS/B73 mutant; lines 2 and 3 B73 cat transgenic line; B73 as control.

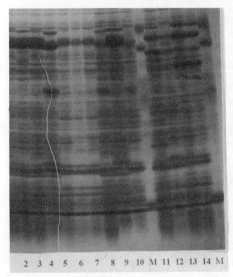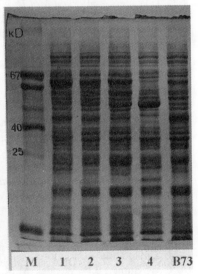

Plate II. A. Effect of *A. tumefaciens* DNA on soluble protein alteration in dry embryo: lines1 - 9, inbred Ms71A x B6S3; line 10, Ms71A control; line 11, Mo17 x NPT II; line 12, Mo17 x cat DNA; line 13, Mo17 X NPT II ; line 14, Mo17control; and line M, m.w.markers.
B. Effect of NPT II gene on soluble protein alteration in dry embryo: line M, m.w.marker; lines 1 - 4, transgenic B73 inbreds and last line B73 as control.

4   CONCLUSIONS

There is evidence presented on bacterial DNA integration into maize genome by the dry seed incubation in solution of plasmide pRT100*neo* - neomycin phospho transferase DNA, *A.tumefaciens* DNA and *E.coli* plasmide containing *cat* sequence DNA. After foreign gene integration, useful mutations have been selected and tested in standard genetics experiments on inheritance of induced mutation (earlier flowering and maturation of transgenic lines e.g.) and performance of combining ability, as important traits from a breeding point of view.

References

1. P. Zambryski, L. Herrera-Estrella, M. De Block, M. van Montagu and J. Schell, "Genetic Engineering, Principles and Methods", Plenum Press New York, 1984, p.53.
2. J. Schell ,*Genetika*, 1987, **19**, p.129.
3. I. Potrycus, *Physiol. Plant* , 1990, **79**, p.125.
4. L.E. Murry et al. *Bio/Technology* , 1993, **11**, p.1559.
5. C.M. Buising R.M.Benbow, *Mol.Gen.Genetics*, 1994, **243**, p.71.
6 C.M. Laursen et al. *Plant Mol. Biol.*, 1994, **243**, p.51.
7. K.D. Halluin et al. *Plant cell*, **4**, p.1495.
8. D.A. Walters et al. *Plant Mol.Biol*, 1992, **18**, 2, p.189.
9. M.C. Lusardi et al. *Plant Journal*, 1994, **5**, 4, p.571.
10. K. Konstantinov and M. Denić, Second International Congres of Plant Mol. Biol. 1988, Jerusalem, p.60.
11."Molecular cloning: a laboratory manual" First Edition, Cold Spring Laboratory Press 1982.
12. M. De Block, L. Herrera Estrella, M. van Montagu, J. Schell and P. Zambrisky, *EMBO. J.*, 1984, **3**, p.1681.
13. U. Wienand, A. Schrell, H. Saedler, "Maize 90. Maize production, processing and marketing in mediteranien countries", Maize Res. Inst. Belgrade 1990, p.157.
14. J. Southern, *J.Mol.Biol.*, 1975, **98**, p.503.
15. Y. Ohta, *Proc. Natl. Acad. Sci USA*, 1986, **83**, p.715.
16. T. Murashige. and F.Skoog, *Physiol.plantarum.*, 1982, **15**. p.473.
17. S. Mladenović, K. Konstantinov, M. Denić and B. Tadić, *Genetika*, 1991, **23**, p.191.
18. G. Van den Brock, M.P. Timko, A.P. Kaisch, A.R. Cashmore, M. Van Montagu and L. Herrera-Estrella, 1985, *Nature*, **313**, p.358.
19. L. Herrera-Estrela, G.Van den Brock, R. Maenhaut, M. Van Montagu, J. Schell, M. Timko and A. Cashmore, *Nature*, 1984, **310**, p.115.
20. C.W. Stuber, J.F. Wendel, M.M. Goodman, J.S.S. Smith,1988, N.C. State Expt. Res. Bull. 1988, No. **286**, p.87.
21. M. Zivy. H. Thiellement, D. de Vienne, J.P. Hofmann, *Theor. Appl. Genet.*, 1983 **66**. p.1.
22. C. Wang, K. Bian, H. Zhang, Z. Zhou, J. Wang, *Seed Sci.and Technology*, 1994, **122**, p.51.
23. U.K. Laemly, *Nature*, 1970, **227**, p.680.
24. P. Aligoy and J. Duesing, *Bio/Technology*, 1995, **13**, p.454.

25. M.T. Spencer et al. 1992, *Plant Mol. Biol.,* 1992, **18**, p201.
26. W. Schuh et al., *Plant Sci.*, 1993, **89**, p.69.
27. A. De la Pena, H. Lorz, J. Schell, *Nature*, 1987, **325**, 274.
28. K. Konstantinov, S. Mladenović, M. Denić, *Genetika*, 1991, **23**, 2, p.121.
29. S. Mladenović, K. Konstantinov, M. Denić, B. Tadić, *Genetika,* 1991, **23**, 3, p.191.
30. K. Konstantinov, S. Mladenović, G. Saratlić, S. Gošić, B. Tadić, V. Sukalović, J. Dumanović, "Breeding and molecular Biology:Accomplishments and future promises", Proc. XVIth EUCARPIA Conference. 1993, Bergamo, p.44.
31. A.C.F. Graves and S.L. Goldman, *Plant Mol.Biol.,* 1986, **7**, 43.
32. N. Grimsley, T. Hohn, J.W. Davies and B. Hohn, *Nature*, 1987, **325,** 177.
33. R.A. Guse, J.G. Coors, P.N. Drolsom, W.F. Tracy, *Theor. Appl. Genet.,* 1988, **76**, p.398.
34. J.F. Doebley, M.M. Goodman, C.W. Stuber, *Am. J. Bot.,* 1983, **72**, p.629.
36. M.L. Abler and J.G. Scandalios, *Theor. Appl. Genet.*, 1991, **81**, p.635.
35. P. Longo and D. Scandalios, *Proc. Natl. Acad. Sci USA.* , 1969, **62**, p.104.
37. M.E. Vayda and M. Freeling, *Plant Mol. Biol.,* 1986, **6**, p.441.
38. N.S. Yang and D. Scandalios, *Arch. Biochem. Biophys.*, Vol 171,p.575.
39. W.J.G. Gordon-Kamm et al. *Plant cell*, 1990, **2**, p.603.
40. Konstantinov et al. *Maize Genetics News Lett.,* 1996, **70**, p.74.

PCR-based Systems for Evaluation of Relationships Among Maize Inbreds

A. Rafalski, M. Gidzińska and I. Wiśniewska

PLANT BREEDING AND ACCLIMATIZATION INSTITUTE, DEPARTMENT OF TISSUE CULTURE, RADZIKOW, 00-950 WARSZAWA, POB 1019, POLAND

INTRODUCTION

Since the first publication on the use of PCR-based RAPD marker system[1] and on PCR-based DNA fingerprinting[2], several modifications of these methods have been developed. The modification of systems for marker studies were generally directed to conversion of dominant RAPD markers into codominants and resulted in elaboration of such as systems as SCAR or AFLP. The modifications of the DNA fingerprinting system were focused on the development of more complex, polymorphic and reproducible band patterns.

Although the present DNA fingerprinting systems such as DAF and AP-PCR generate very complex fingerprints and are much more informative than the RAPD system they are also more time and cost consuming because of more complicated systems of band separation and visualization (silver staining or radioactive detection)[3]. Thus, the RAPD system is the most commonly used, despite its low resolving power and the fact that part of the RAPD primers have to produce noninformative or low-informative band patterns. The increase in information on DNA sequences of plant genomes gives the possibility to apply selected primers which target specific regions of plant genome. For particular plant species, systems based on some gene families or dispersed repetitive sequences have been developed.

RESULTS

In our studies we have developed a system based on intron splice junction sequences, proposed first by Weining and Langridge[4]. This system seems to be universal for plants because is based on consensus sequences common for plants and necessary for effective splicing[5] (Fig. 1). The use of these sequences as the target for PCR primers gives the possibility to construct four series of semirandom primers. The 5' site of upper strain sequence and complementary 9 base sequence of lower strain were supplemented with 9 random bases to give 18 base exon targeting or intron targeting primers. The additional bases were added to extend primer length and for generating variability of primers. At the 3' site of the upper strain, part of the longer (16 base) sequence consists of a poli A/T tract inconvenient for use as primer. Thus, only 7 bases adjacent to exon were treated as target sequence. These sequences were supplemented with 3 bases at random. These shorter (10 base) primers have annealing characters similar to those of RAPD primers and may be combined with them.

The generation of band patterns by (ISJ) semirandom primers and under the conditions of RAPD system gives the possibility to compare the utility of both systems for evaluation of genetic distance between maize inbreds.

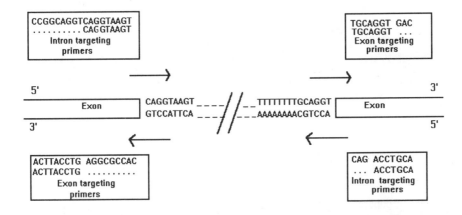

Figure 1. *The location of targets for intron splice junction (ISJ) primers.*

The twelve maize inbreds were chosen from breeding program of Dr. R.Warzecha of our Institute. They represent both flint and dent type, originate from different sources but evaluation of they pedigree is in most cases very difficult. The main character of inbreds is the suitability for the development of early hybrids.

Some RAPD primers generated quite complex and polymorphic band patterns (Fot. 1), but this is not a rule and noninformative or low-informative band patterns were generated by other primers (Fot.2).

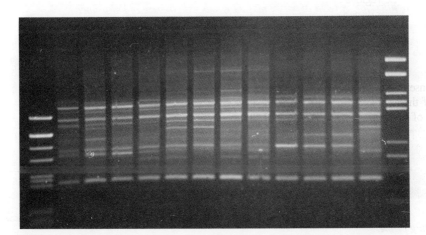

Fot.1. *Amplification of maize DNA using RAPD primer (informative results).*

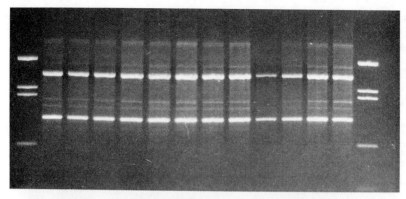

Fot.2. *Amplification of maize DNA using noninformative RAPD primer.*

In our experiments with the use of ISJ semirandom primers we have not obtained the results which can be evaluated as noninformative. In the preliminary experiments, the complexity of patterns had to be tailored by increases in annealing temperature (Fot.3).

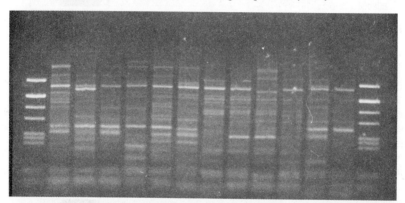

Fot.3. *Amplification of DNA of twelve maize inbreds with the use of ISJ primer.*

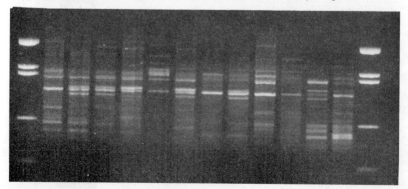

Fot.4 . *Amplification of DNA of twelve maize inbreds with the use of an 18 base ISJ, exon targeting primer.*

The 18 base ISJ primers used alone generated complex and highly polymorphic band patterns and usually the polymorphisms between all inbreds were detected. We have not found any essential difference in the action of exon targeting and intron targeting primers (Fot.4,5).

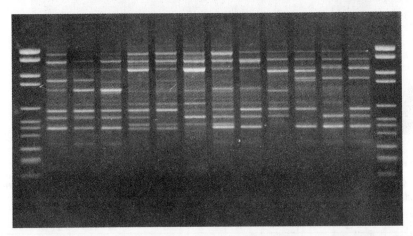

Fot.5. *Amplification of DNA of twelve maize inbreds with the use of an 18 base ISJ, intron targeting primer.*

The short 10 nucleotide ISJ primers can be used in combination with RAPD primers, especially those, when used alone, amplified noninformative or low-informative band patterns (Fot.6,7).
The indexes of similarity (according to Nei & Li)[6] calculated from experiments with the use of RAPD primers (Table 1) varied from 0.71 to 0.94 with the average value of 0.81.
The indexes of similarity of the same 12 inbreds calculated from semirandom experiments varied from 0.43 to 0.80. (Table 2) and the average value was 0.62. This indicates that with the use of ISJ primers we are able to detect polymorphism even between closely related genotypes.

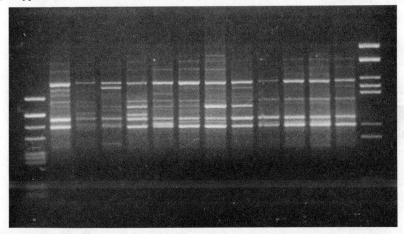

Fot. 6. *Amplification of DNA of twelve maize inbreds using RAPD primer (Rad.24).*

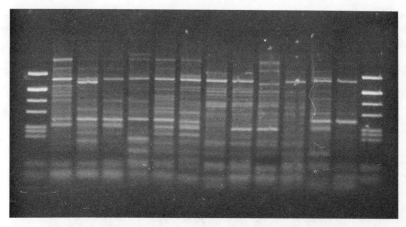

Fot.7. *Amplification of DNA of twelve maize inbreds with the use of a 10 base ISJ primer coupled with RAPD (Rad .24) primer.*

	2	3	4	5	6	7	8	9	10	11	12
1	0.84	0.77	0.79	0.84	0.80	0.74	0.80	0.80	0.71	0.79	0.75
2		0.77	0.78	0.79	0.79	0.76	0.80	0.81	0.77	0.83	0.81
3			0.78	0.76	0.74	0.73	0.79	0.76	0.80	0.76	0.72
4				0.76	0.76	0.83	0.71	0.82	0.75	0.74	0.72
5					0.94	0.85	0.80	0.86	0.80	0.84	0.83
6						0.85	0.84	0.89	0.83	0.86	0.86
7							0.79	0.79	0.84	0.82	0.80
8								0.86	0.81	0.84	0.79
9									0.83	0.84	0.84
10										0.81	0.87
11											0.83

Table 1. *Indexes of similarity calculated from experiments with the use of seven RAPD primers.*

	2	3	4	5	6	7	8	9	10	11	12
1	0.62	0.61	0.61	0.62	0.65	0.56	0.60	0.64	0.52	0.62	0.61
2		0.52	0.57	0.72	0.66	0.60	0.57	0.56	0.56	0.69	0.55
3			0.72	0.52	0.51	0.59	0.63	0.64	0.63	0.57	0.53
4				0.56	0.47	0.59	0.74	0.73	0.73	0.71	0.66
5					0.61	0.62	0.64	0.73	0.54	0.67	0.58
6						0.50	0.59	0.46	0.42	0.59	0.55
7							0.61	0.66	0.67	0.69	0.59
8								0.63	0.64	0.62	0.63
9									0.65	0.71	0.61
10										0.75	0.65
11											0.80

Table 2. *Indexes of similarity calculated from experiments with the use of six semirandom primers.*

CONCLUSIONS.

In comparison to RAPD system, ISJ - based semirandom primers generate slightly more complex and much more polymorphic band patterns.

The approach with the use of ISJ consensus sequences connected with some random sequences can be developed without any essential limitation.

The use of a limited number of ISJ primers, alone or in combination with random primers, decreases the costs of PCR analysis.

The analysis with the use of ISJ primers reveals polymorphism mainly in the transcribed regions of plant genome.

REFERENCES.

1. J.G.K. Williams, A.R. Kubelik, J. Livak, J.A. Rafalski, C.V. Tingey, *Nucl. Acid Res.,* 1990, **18**, 6531-6535.
2. J. Welsh, M. McClelland, *Nucl. Acids Res.,* 1990, **19**, 861-866.
3. G. Caetano-Anolles, *Plant Mol. Biol.,* 1994, **25**, 1011-1026.
4. S. Weining, P. Langridge, *Theor. Appl. Genet.,* 1991, **82**, 209-216.
5. J.W.S. Brown, G. Feix, D. Frendewey, *EMBO Journal,* 1986, 2749-2758.
6. N. Nei, W.H. Li, *Proc. Natl. Acad. Sci USA,* 1979, **76**, 5269-5273.

SESSION IV
THE ROLE OF ENVIRONMENTAL CONDITIONS IN
MAIZE AND SORGHUM DEVELOPMENT & BREEDING

Overcoming Inbred Line Stagnation for Productivity and Stability in Maize Breeding

A. C. Fasoulas

DEPARTMENT OF GENETICS AND PLANT BREEDING, ARISTOTELIAN UNIVERSITY OF THESSALONIKI, 540 06 THESSALONIKI, GREECE

1 INTRODUCTION

Duvick[1] in a noteworthy message springing from his personal history, remarked: "Corn breeding has made startling advances in speed and accuracy through improvements in mechanical, electronic and communications technology. But breeding techniques today would be immediately recognizable to a breeder from the 1930s. Producing single-cross rather than double-cross hybrids is the only real change, the only fundamental modification in breeding strategy. With all our advances in maize genetics and technology, we still do not know what causes hybrid vigor, nor do we know the genetics of yield or other important agronomic traits such as drought tolerance or resistance to root lodging. Cut-and-try is still the best genetic procedure in corn breeding". According to Duvick[1]: "This dichotomy, technological progress versus genetic stagnation, is true for other crops, as well. Breeders universally depend on experience and art more than on genetics".

The objective of the present paper is to explore the causes of maize inbred line stagnation for productivity and stability, in an effort to overcome stagnation by developing more efficient breeding strategies. For details concerning the scientific principles underlying the new breeding strategies the reader is referred to volumes 13, 14 and 15 of Plant Breeding Reviews.[2-4]

Description and causes of stagnation

Inbred line stagnation for productivity and stability is first reflected by the level of the yield of inbred lines relative to the yield of single crosses that remained below 50% for more than 50 years;[5] second, by the inability to predict hybrid performance. The cause of stagnation is the accumulation of defective genes in the genomes of inbred lines on account of three uncritical choices.

1.1 First Uncritical Choice

The first uncritical choice is the use of field plots to mimic the crop environment ignoring the effects of density and competition, (1) on single plant heritability, and (2) on crop productivity.[3,4]

1.1.1 Effects of density and competition on single plant heritability. Figure 1 illustrates how the three basic parameters describing a population, i.e., the mean ($\bar{X}$),

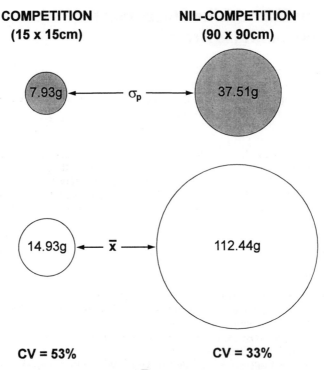

Figure 1   *The fate of the mean ($\overline{X}$), σ_p and CV in a rye population grown under competition and nil-competition (source[4]).*

the phenotypic standard deviation (σ_p), and the coefficient of variation (CV), are affected by density when two samples of the same rye population were grown side by side in the presence and absence of competition. In Figure 1 three events merit special attention: (1) Maximal phenotypic differentiation, indispensable for efficient selection, is accomplished in the absence of competition where σ_p and $\overline{X}$ get maximal values; (2) as $\overline{X}$ increases at a higher rate than σ_p, it quantifies better phenotypic differentiation; (3) the increase of CV under competition and its reduction at nil-competition, means that competition brings about more undesirable phenotypic variation compared with nil-competition.

   The impact of the aforementioned events on single plant heritability is decisive. First, single plant heritability is optimized in the absence of competition where phenotypic expression and differentiation as quantified by $\overline{X}$, are maximized. Second, reduction of CV in the absence of competition, improves heritability by increasing the share of genetic variance at the expense of environmental variance in the denominator of the equation:

$$h^2 = \sigma^2_g / \sigma^2_g + \sigma^2_e + \sigma^2_{gxe}$$

Third, the negative correlation between yielding and competitive ability[6] erases the correspondence between genotypic and phenotypic values and nullifies heritability. Moreover, it brings about variety degeneration as long as varieties are reproduced under dense stand, due to the differential survival of strong competitors-low yielders at the expense of weak competitors-high yielders.

The outcome from the aforementioned findings is that efficient selection for yield on a single plant basis is possible only in the isolation environment. This is verified by the data presented in Table 1 where it is shown that genetic advance through mass selection is inversely proportional to density. The data of Table 1 explain why efforts to select for yield on a single plant basis in the crop environment, have been unsuccessful. In essence, the results of Table 1 impose replacement of conventional field plots as units of evaluation and selection, by single plant plots grown in the absence of competition. In other words, they establish nil-competition as the sole condition for evaluation and selection at all stages of the breeding program.

1.1.2 Effects of density and competition on crop productivity. Fasoula and Fasoula[3] have described the condition under which crop productivity is maximized as the one which ensures that densely grown plants share equally the limited growth resources, so that the yield of the competing plants is evenly suppressed. This condition is termed **positive competition**, to be distinguished from **negative competition**, which involves unequal sharing of resources and reduction of crop productivity. Crop productivity is reduced under negative competition as gains from competition fail to balance the losses. Negative competition is quantified by the magnitude of CV of single plant yields. CV=33% determines the limit above which productivity is reduced as competition turns into negative, and below which productivity is increased as competition turns into positive.

1.1.3 Prerequisites for positive competition and maximal productivity. The necessary conditions for equal sharing of the limited growth resources and therefore for maximal productivity, are the following: first, growth resources must be ample and evenly distributed; second, in order to exclude unequal sharing due to genetic differences among plants, cultivars must be monogenotypic (hybrids, pure lines, clones); third, in order to exclude unequal sharing due to acquired differences among plants, growth and development of the genetically identical plants must be synchronous.

The third precondition needs further elaboration. To eliminate acquired differences among plants, apart from the measures ensuring an even distribution of

Table 1 *Genetic Advance through Mass Grid and Honeycomb Selection is Inversely Proportional to Density. This is Because of the Confounding Effects of Density and Competition on Single Plant Heritability.*

Reference	Sampled Populations	Planting Rate (plants/m^2)	Response %
A.R.Hallauer and J.H.Sears, *Crop Sci.*, 1969, **9**, 47.	Krug and Iowa Ideal	3.9	0.5-1.5
C.O.Gardner, *Crop Sci.*, 1961, **1**, 241.	Hays Golden	1.8	3.9
A.H.A.Onenanyoli and A.C.Fasoulas, *Euphytica*, 1989, **40**, 43.	Pioneer 3183(F$_2$)	0.7	11.23

growth resources and the synchronization of germination and growth, the most decisive role is played by the so called tolerance to density. This is a genetic property that has to do with the incorporation of genes conferring resistance to the stress of density. Tolerance to density eliminates negative competition by eliminating acquired differences among plants, allowing thus use of high plant densities and optimization of productivity. As it will be explained shortly, tolerance to density by reflecting the sum of resistances to various biotic and abiotic stresses encountered in the target environments, can be anticipated from evaluation in the isolation environment. In other words, genes responsible for tolerance to density may be incorporated by selecting for yield potential and yield stability in the isolation environment.

1.2 Second Uncritical Choice

The second uncritical choice was the rank of entries built exclusively on the basis of the mean, leaving unexploited the ranking potentialities of phenotypic standard deviation. As shown by Fasoula and Fasoula[3], phenotypic standard deviation of entries grown in the isolation environment reflects the load of defective genes that affect stability of performance. The larger the phenotypic standard deviation in the isolation environment, as expressed by the magnitude of the coefficient of variation (CV), the higher the load of defective genes and therefore the more reduced the buffering and tolerance to density. This finding opened the way to develop criteria that anticipate stability of performance and tolerance to density in the crop environment, by using the value of phenotypic standard deviation in the isolation environment. In fact, research in recent years[7-9] has shown that the yield improvement of modern cultivars resulted more from the improvement of yield stability than from the improvement of yield potential.

That yield stability is more important than yield potential is shown in Figure 2 where the yield potential of five distinct categories of tomato cultivars is compared under various plant densities. It is interesting to notice that under high plant density all cultivars yielded the same. This means that once genes for tolerance to density are incorporated into the cultivars, reduced yield potential may be compensated for by the use of higher planting rates.

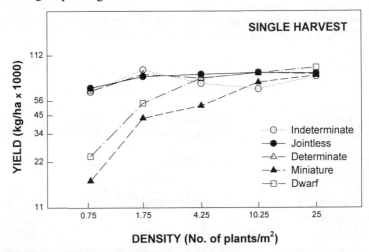

Figure 2 *Effect of density on fruit yield differentiation in five tomato cultivars (source[10]).*

The finding that in the isolation environment, $\overline{X}$ quantifies the yield potential of entries, and σ_p the yield stability, led Fasoula and Fasoula[4] to develop the prediction criterion:

$$PC=\overline{X}(\overline{X}_s-\overline{X})/\sigma_p^2$$

where $\overline{X}_s$ the mean yield of plants selected by truncation within each entry by a chosen selection pressure. The prediction criterion is scale independent and allows to anticipate yield potential and stability in the crop environment, based on evaluation in the isolation environment. It allows further, to select two distinct categories of entries, those combining high yield stability with high yield potential, and those combining high yield stability with low yield potential. The validity of the prediction criterion depends on the validity of the estimation of $\overline{X}$ and σ_p; in other words, on the use of the honeycomb selection designs[2] that allow reliable sampling for environmental diversity, and objective selection among entries. As shown in Figure 3, the establishment of honeycomb trials is very easy because entries are laid in horizontal rows, in an ascending order left to right. Figure 4 shows how the honeycomb designs allocate entries under comparable growing conditions. This is accomplished by laying entries in an equilateral triangular lattice (ETL) pattern that samples for environmental diversity more effectively than random allocation.

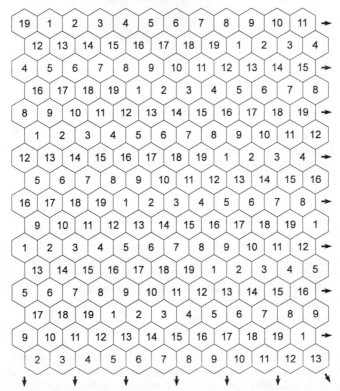

Figure 3 *The replicated-19 honeycomb design evaluates a maximum number of 19 entries allocated in horizontal rows in an ascending order left to right.*

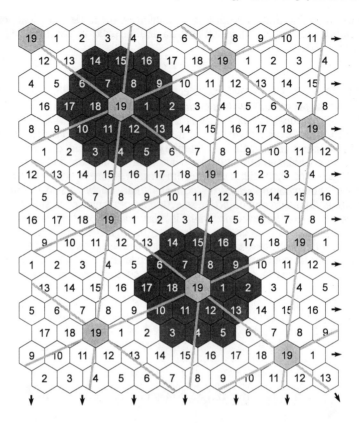

Figure 4 *Allocation of entries in a equilateral triangular lattice (ETL) pattern, enables effective selection among entries due to the effective sampling for environmental diversity. Selection within entries is accomplished through moving replicates (shaded area).*

Evidently, entries replicated many times in an ETL pattern across the target environments, are laid under comparable growing conditions. Therefore, evaluation and selection of entries on the basis of PC across production environments, allows selection of entries exhibiting superior performance when grown at high planting rates. Honeycomb designs[2] also ensure objective selection within entries by coping effectively with the disturbing effects of spatial heterogeneity on single plant yields. This goal is accomplished by means of moving replicates of a quasi-circular shape. According to the principle of the moving replicate, all plants belonging to a given entry lie in the center of concentric rings in the periphery of which occur plants belonging to the remainder entries (Figure 4). This gives the possibility to convert the yield of the central plant in percent of the yield of a multi-entry common check and erase the disturbing effects of spatial heterogeneity. Moving replicates allow also to adjust objectively selection pressure within entries by using as a common check the average of entries occurring within a concentric ring of a chosen size. For example, using as a check the average of entries occurring within the first, the second, the third and so on, ring, selection intensity increases accordingly, and by extent response to selection.

The use of the prediction criterion (PC), which allows to anticipate performance in the crop environment based on evaluation in the isolation environment, will be issustrated with data[11] presented in Table 2. Table 2 is divided into three sections. In each section are given the entry code, the average yield per plant ($\overline{X}$), and the prediction criterion values (PCV). The data in the first section concern S_6 lines derived from the F_2 of the hybrid LORENA (PR-3183) through controlled selfing. The data of the second section concern S_6 lines extracted from the same F_2 after previous population improvement by six cycles of honeycomb selection[12]. The third section comprises random hybrids between the two sets of S_6 lines. All entries have been tested in the isolation environment in 1996 in a honeycomb field trial and are ranked in a descending order according to the yield potential as reflected by the average yield per plant ($\overline{X}$). On the basis of yield potential, lines derived after previous population improvement, are on the average superior to lines derived directly from the F_2. Also, in the isolation environment, LORENA had the lower yield potential, although its yield potential in the crop environment was insurmountable, as indicated by the fact that for more than 10 years LORENA was the leading hybrid in Greece. This means that yield potential in the isolation environment, does not ensure reliable evaluation. By contrast, when entries are ranked on the basis of the prediciton criterion (PC), which evaluates jointly for yield potential and yield stability, LORENA gets the first position (PCV=10.9). When entries are ranked on the basis of PCV, lines derived directly from the F_2 are superior to lines derived following previous population improvement. This means that prolonging population improvement at nil-competition in the hope to improve yield potential, may lead to a gradual loss of stability genes and to a loss of a real genetic progress. This is especially true for selection performed in a single environment.

Table 2 *Prediction Criterion Values (PCV) in Two Sets of S_6 Lines and in Random Hybrids Among Them, Grown in the Absence of Competition (125 cm).*

S_6	$\overline{X}$ (g)	PCV	S_6	$\overline{X}$ (g)	PCV	Hybrid	$\overline{X}$ (g)	PCV
35*	657	3.3	48	507	3.6	23*x24*	1437	7.3
35	361	6.0	20*	472	5.0	5x6	1395	7.2
41*	353	3.5	18*	465	4.5	9x8	1315	5.1
47	326	8.0	48*	460	4.6	21*x20*	1282	5.4
11	314	6.2	6*	454	3.8	55*x57*	1281	5.2
5	305	5.9	24*	450	4.6	9*x8*	1278	4.1
53	305	5.0	6	445	5.4	53*x54*	1275	6.9
.	.	.	.	.	.	.	.	.
.	.	.	.	.	.	.	.	.
21	242	6.3	36*	309	3.7	35*x36*	1007	5.8
3	228	8.3	38	304	3.4	11*x12*	1005	3.6
11*	225	3.8	30	290	5.3	53x54	998	6.2
45	218	9.0	50*	287	5.0	58x60	985	4.5
17	217	3.7	57	279	2.9	45x44	905	4.6
41	198	5.4	18	263	3.8	3x2	904	4.9
15	190	4.3	38*	226	2.1	17x18	673	4.3
						LORENA	668	10.9

1.3 Third Uncritical Choice

The third uncritical choise concerns the marked preference for hybrid over inbred line cultivars. As long as the genetic basis of heterosis remains an unsolved mystery, this preference is arbitrary and unjustifiable. The solution to the mystery of heterosis is related to the elucidation of whether or not there is true overdominance. According to various investigators,[13,14] a definitive distinction between true and pseudo-overdominance will require fine mapping of QTLs displaying overdominant gene action. In the past, fine maping of overdominant loci was impractical. Now, with the availability of high-density molecular linkage maps, fine mapping is a feasible proposition and the hypothesis that overdominant QTLs in maize are the result of tight linkage of dominant and recessive alleles can be tested empirically.

A broad generalization[15] covering all studies of the various types of gene action in maize, indicates the almost universal importance of additive effects. Nonadditive effects may exist, but are commonly smaller in magnitude. Eberhart[16] pointed out that statistical genetic studies suggest that complete dominance may be much more important than overdominance in the improvement of maize hybrids. In general, empirical population improvement studies with maize have failed to support the hypothesis of overdominance. Jinks,[17] investigating the genetical basis of current plant breeding strategies, as typified by the production of F_1 hybrids at one extreme and by recombinant inbred lines at the other, arrived at conclusions that may be summarized as follows: true overdominance is not a cause of F_1 heterosis. This means that recombinant inbred lines superior to existing inbred lines and to their F_1 hybrids may be recovered with the predicted frequencies. Therefore, the choice between hybrids and recombinant inbred lines as the end-product must rest on other non-genetic considerations.

Figure 5 gives in a schematic form the accumulated evidence on the control of heterosis.[4] As indicated in Figure 5, intragenic crossing over may convert unfixable pseudo-overdominant into fixable dominant effects. It is indicated also, that pseudo-overdominance and dominance have a common characteristic, i.e., they are coupled with genetic defects. This means that selection schemes aiming at exploiting pseudo-overdominant and dominant gene action, lead inevitably to the preservation and perpetuation of genetic defects. It also means, that the presence of defective genes impairs drastically not only inbred line productivity and stability, but also the ability to anticipate hybrid performance. This is because once an inbred has been isolated, the chances of finding a second inbred covering or restoring mutually allelic defects, are very limited.

Thus, heterosis by basing its superiority on the accidental covering and restoring of allelic defects, sets insurmountable barriers. First, it promotes accumulation of defective genes. Second, it hinders prediction of hybrid performance. Third, it impairs single plant heritability. Fourth, it limits the chances to exploit additive alleles.

The tight linkage of heterosis with allelic defects, imposes radical changes in maize breeding strategy which consist of abandoning combining ability as a selection criterion, and focusing exclusively on inbred line performance *per se.*

1.3.1 Key steps to efficient selection for inbred line performance per se. Inbred line performance *per se* in the isolation environment involves three main objectives: (1) yield potential, (2) yield stability, and (3) quality. Yield stability is the most important objective and the most difficult to improve as it involves incorporation of genes conferring resistance to several predictable and unpredictable biotic and abiotic stresses encountered across the target environments.

The key steps to efficient selection for inbred line performance *per se* in the isolation environment, may be summarized as follows:

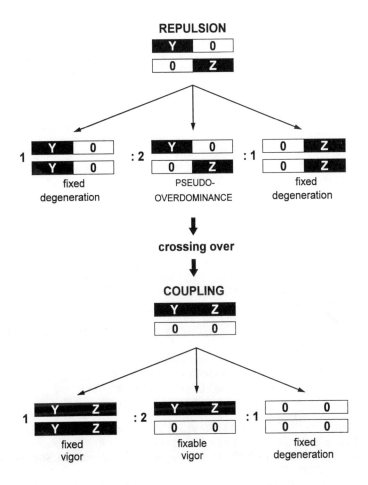

Figure 5   *Schematic presentation of the role of pseudo-overdominance and dominance in the control of heterosis. Intralocus crossing over may convert unfixable pseudo-overdominant heterotic effects, into fixable dominant effects (source[4]).*

1. Treat maize as a self pollinator by applying controlled selfing to fully exploit additive genes.
2. Sample jointly for productivity, stability, and quality across production environments.
3. Utilize the honeycomb designs to accomplish two important goals: (a) sample for environmental diversity, to select effectively between entries, and (b) cope with the confounding effects of spatial heterogeneity on single plant yields, to select effectively within entries.
4. Utilize the prediction criterion (PC) to anticipate yield and tolerance to density in the crop environment, based on evaluation in the isolation environment.

The aforementioned strategy was applied[18] to the F_2 of the single cross maize hybrid LORENA (PR-3183). The result was the development of very productive inbreds, one of which (D-17) represented 70% of the LORENA potential. This is much above the 50% potential of current elite inbreds. This shows clearly that inbreds with 90% of the single cross potential can be easily produced. One consequence from improving inbred line performance *per se*, will be the reduction of heterosis. For example, using inbreds with 50% of the single cross potential, heterosis will be 100%. For 70% of the single cross potential, heterosis is reduced to 43%, and for 90% potential to 11%. Reduction of heterosis is accompanied by a number of advantages: (1) inbred lines may easily reach the yield potential and stability of hybrids; (2) the very limited number of outstanding inbreds, renders useless the need to predict hybrid performance; (3) the high productivity of inbreds reduces drastically the cost of hybrid seed; (4) the high productivity of inbreds carries along the productivity of hybrids on account of the replacement of defective by favorable additive alleles following breakage of repulsion linkages.

References

1.   D.N.Duvick, *Crop Sci.*, 1996, **36**, 539.
2.   A.C.Fasoulas and V.A.Fasoula, 'Plant Breed.Rev.', 1995, **13**, 87.
3.   D.A.Fasoula and V.A.Fasoula, 'Plant Breed.Rev.', 1997, **14**, 89.
4.   D.A.Fasoula and V.A.Fasoula, 'Plant Breed.Rev.', **15** (in press).
5.   D.N.Duvick, (W.R.Fehr, ed), Spec. Publ. No.7, *Crop Sci.Soc.Am.*, Madison, Wisconsin, 1984, **7**, 15.
6.   D.A.Fasoula, *Euphytica*, 1990, **50**, 57.
7.   M.Derieux, M.Darrigand, A.Gallais, Y.Barriere, O.Bloc and Y.Montalant, *Agronomie*, 1987, **7**, 1.
8.   J.Bingham, 'Phil.Trans.R.Soc.Lond.' 1981, **B 292**, 441.
9.   W.A.Russell, 'Advances in Agronomy', 1991, **46**, 245.
10.   R.L.Fery and J.Janick, J.Am.Soc.Hort.Sci.,1970, **95**, 614.
11.   J.S.Tokatlidis, Ph.D. Thesis, University of Thessaloniki, 1997.
12.   A.H.A.Onenanyoli and A.C.Fasoulas, *Euphytica*, 1989, **40**, 43.
13.   A.H.Paterson, E.S.Lander, J.D.Hewitt et al., *Nature*, 1988, **335**, 721.
14.   C.J.Jansen and P.Stam, *Genetics*, **136**, 1447.
15.   G.F.Sprague and S.A.Eberhart, (G.F.Sprague ed.), American Society of Agronomy, Madison, Wisconsin, 1977.
16.   S.A.Eberhart, (E.Pollak, O.Kempthorne and T.B.Bailey, Jr eds), 'Proc. Int. Conf. Quantitative Genetics', 491.
17.   J.L.Jinks, 'Phil.Trans.R.Soc.Lond.' 1981, **B 292**, 407.
18.   C.G.Ipsilandis, Ph.D.Thesis, University of Thessaloniki, 1996.

Epigenetic Modifications of Total Genomic Maize DNA: The Role of Growth Conditions

A. S. Tsaftaris, M. Kafka and A. Polidoros

DEPARTMENT OF GENETICS AND PLANT BREEDING, ARISTOTELIAN UNIVERSITY OF
THESSALONIKI, 540 06 THESSALONIKI, GREECE

1 INTRODUCTION

From studies involving estimations of polymorphisms either at the protein level (Protein Amount Polymorphism) (1) and the mRNA level (RNA Amount Polymorphism) (2) in different maize genotypes the significance of regulatory mechanisms for the quantitative regulation of gene expression was stressed. One such mechanism involved in regulating the amount of expression output of maize genes is methylation of cytosine residues in its DNA. The extent and distribution of genomic DNA methylation was found to be significantly correlated with the rate of expression of many genes examined, not only in plants, but in all higher organisms as well (3).

The goal of this work was to study the extent of cytosine methylation in maize genomic DNA, its variation among different genotypes and developmental stages. The role of maize growth conditions in the field on the extent of genomic DNA methylation was also examined.

DNA methylation was measured following different approaches and at different levels, namely as: percent methylation of the whole genomic DNA, methylation of specific gene families such as rDNA genes, methylation of random regions of the genome and finally of a specific individual gene.

We report here the results of the first approach; that is measuring percent methylation of whole genomic DNA in spaced and dense grown maize plants in the field and in different developmental stages of many different genotypes. Significant differences were found in the percent of total DNA methylation between spaced and dense grown maize plants. Furthermore, methylation was found to be developmental stage and genotype specific.

2 MATERIALS AND METHODS

2.1 Genetic material

The genetic material analysed in the work presented here consisted of seven maize inbreds NE2, 5C, 102A, 22M, 30B, 60M, 4D and two of their hybrids 60MX30B, 22MX102A.

Abbreviations: HPLC, High Performance Liquid Chromatography; Adenine A;Thymine T; Guanine G; Cytosine C; 5-methylcytosine 5-mC; Polymerase Chain Reaction, PCR.

The inbred NE2 derived from the Indiana lines ITE 701X07, whereas of inbred 5C was produced after 8 generations of self-pollination of the F68XNE2 hybrid which is known as IS027 hybrid of the Greek Cereal Institute. The other inbreds are S_4 lines derived from the F_2 generation of the single hybrid Lorena after 5 years of mass honeycomb and pedigree selection (at the fifth year). At the C_5 selection cycle the more rigorous plants were self-pollinated and the above inbred lines and hybrids were created (4).

2.2 Field experimentation

The plant material used in this work was grown at the Experimental Farm of Aristotle, University of Thessaloniki using the honeycomb experimental design (5). Maize plants were sown in two different plant densities: 1.5m (spaced) and 25cm (dense) distances between individual plants with a density of 0.513plants/m^2 and 18.5plants/m^2, respectively.

2.3 Genomic DNA isolation

DNA was extracted as described (6) with minor modifications, from leaves of plants grown in the field for 40 (stage 1) and 60 (stage 2) days in the two different densities. All DNA samples were treated with RNase-DNase free (10mg/ml) to remove RNA. The integrity of DNA samples was examined by agarose gel electrophoresis and quantified by measuring the optical density in a spectrophotometer at 260nm.

2.4 Percent genomic DNA methylation using isoschizomeric restriction enzymes

Two different pairs of isoschizomeres were used: the MspI - HpaII and the BstNI - EcoRII. The first pair recognizes the CCGG sequence of DNA, and while the MspI is insensitive to cytosine methylation, HpaII cuts the CCGG sequence only when the internal cytosine is unmethylated. The digestion conditions were carried out according to manufacturer instructions (GIBCO-BRL) for 3 (MspI) and 12 hours (HpaII).

The second pair recognizes the CC(N)GG [N is Adenine or Thymine] sequence and the BstNI is the insensitive to methylation enzyme, whereas EcoRII is the sensitive one. In plants, methylated cytosines can also be found in the C(N)G sequences (7); that is why the use of the second isoschizomeric pair was taken on account. Digestion was carried out for 1 hour for both enzymes and the conditions used were as instructed by the manufacturer (New England Biolabs).

2.5 Percent genomic DNA methylation using High Performance Liquid Chromatography (HPLC)

DNA, after alkaline hydrolysis to remove all traces of RNA, was hydrolyzed with 99% HCOOH at 170 ^{0}C for 30 minutes; the bases were then lyophilized for 4-5 hours and were separated using HPLC. The molecular concentrations of cytosine and 5-methylcytosine were estimated according to the HPLC as previously described (8) with some modifications, by comparing the heights of the peaks in every chromatogram with those of the standards made with pure bases. All 5 bases (A, T, G, C, 5-mC) were separated according to their different retention times. The results were expressed as percent of methylcytosine (5-mC) to the total (5-mC+C) cytosine of DNA.

In total, 108 different DNA samples were analyzed: 7 genotypes, at 2 developmental stages, grown in 2 different, plant densities, and making 3 replications consisted of different DNA isolations from the same genotype.

3 RESULTS AND DISCUSSION

In Figure 1A the results of restriction enzyme digestion of maize DNA with the methylation sensitive and insensitive isoschizomeres EcoRI and BstNI are shown. In this experiment, total maize DNA from plants growing in the field in spaced or dense conditions was used. First of all, significant differences exist in digesting DNA with methylation insensitive (lanes 2 and 3) in comparison with methylation sensitive enzyme (lanes 4 and 5). The methylation insensitive enzyme completely chops DNA, while DNA treated with the methylation sensitive enzyme remains almost intact. This is a strong argument for maize DNA being a highly methylated DNA, despite the fact that it is difficult to draw conclusions on the exact percent methylation of maize DNA.

As for unravelling the effect of density on DNA methylation following the same approach, by comparing lane 1 with 2 and lane 3 with 4 we can detect differences between them(Figure 1B). Here, the results are shown after using the other isoschizomeric pair of restriction enzymes (MspI and HpaII) at the stage 2 (60 days) for both conditions of maize plant growth. Lanes 1 and 3 are from spaced grown maize plants and lanes 2 and 4 from dense grown plants. As it is shown, DNA in lanes 1 and 3 from spaced grown plants seem to be less methylated and consequently more chopped in relation to lanes 2 and 4 from dense grown plants.

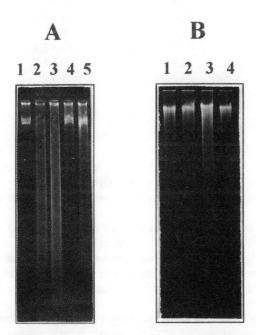

Figure 1 *Restriction enzyme digestion with methylation sensitive and insensitive isoschizomeres*
A. BstNI and EcoRII. Lanes: 1 intact DNA; 2 BstNI digestion spaced; 3 BstNI digestion dense; 4 EcoRII digestion spaced; 5 EcoRII digestion dense
B. HpaII digestion at stage 2 (60 days). Lanes: 1, 3 spaced; 2, 4 dense

In Table 1 the results of total genomic methylation of different maize inbreds in 2 developmental stages measured as percent of total DNA methylated using HPLC, are shown. In this experiment the percent DNA methylation was estimated after acid hydrolysis and separation of DNA bases by HPLC analysis, and it is expressed as the percent of methylated cytosine to the total cytosine of the DNA [5-mC/(5-mC+C) x 100]. At each stage comparisons were also made between percent methylation of maize plants grown in spaced versus dense conditions. The results show that DNA methylation is different for different developmental stages and it is also affected by the conditions of plant growth. At stage 1 where plants are small and they are not affected so much by the density, there are small differences between spaced and dense conditions, but the differences get higher at stage 2 where in the dense condition competition between maize plants has increased.

Table *1* *Percent total genome methylation [5-mC/(5-mC+C) x 100] in maize inbred lines in 2 developmental stages (40 and 60 days) in 2 different plant densities (spaced and dense)*

GENOTYPES	STAGE 1		STAGE 2	
	SPACED	DENSE	SPACED	DENSE
NE2	28.20	28.80	29.30	30.60
5C	30.20	30.30	29.80	30.10
102A	23.10	23.07	24.20	28.71
22M	30.60	27.24	26.22	22.40
30B	21.80	27.70	25.91	27.30
60M	28.30	30.50	27.90	32.87
4D	28.10	28.20	29.47	32.39
Average	27.18	27.97	27.54	29.19

In Table 2 the results of total genomic DNA methylation for the two hybrids are shown. Here, at stage 1 (40 days) the hybrids exhibit higher percent methylation in the spaced planting, but yet at stage 2 (60 days) the percent methylation is higher in the dense condition as well.

Although the difference between spaced and dense condition at stage 2 is higher in comparison to stage 1, still there is only 1.64% difference between spaced and dense for inbred lines and 0.3% for hybrids. But it is well known (9, 10) that minor changes in methylation level could be critical for gene expression, since there is the possibility to have such epigenetic modifications in a region involving or close to the promoter of specific genes.

In accordance with other measurements of percentages of total genomic methylation of maize DNA (2, 11) the average methylation of all our genetic material examined was found to be 27.8% .

Thus our results indicate that the conditions of maize plant growth affect the level of cytosine methylation in plant DNA. Methylation is higher for maize plants grown under stressful (dense) conditions with respect to the total percent methylation of the whole genome as compared with plants grown in optimal conditions (spaced). These indications lead to the suggestion that epigenetic changes could be related to or accompanied with changes to the level of DNA methylation.

Table 2 *Percent total genome methylation [5-mC/(5-mC+C) x 100] in maize hybrids in 2 developmental stages (40 and 60 days) in 2 different plant densities (spaced and dense)*

| | STAGE 1 | | STAGE 2 | |
GENOTYPES	SPACED	DENSE	SPACED	DENSE
60MX30B	31.25	27.32	26.10	26.50
22MX102A	28.62	25.85	26.20	26.40
Average	29.93	26.58	26.15	26.45

Similar results were obtained when the methylation status of random regions of the genome, as well as of specific individual genes, was studied under the two different conditions of plant growth. Methylation of random regions of maize plant genome was studied using enzymatic digestion with methylation sensitive enzymes followed by PCR amplification (Kafka and Tsaftaris, in preparation), and the number of amplified products after HpaII digestion was always higher in the dense condition for all genotypes tested. The methylation status of a specific methylated individual gene, the Ac transposable element, was studied by its phenotypical expression in maize plants grown under spaced and dense conditions. The transpositions and demethylations of the Ac were always higher for the spaced conditions (Kafka and Tsaftaris, in preparation).

These results are of great importance for plant breeding. With conventional plant breeding, breeders always took in to account the genotype and environment interaction, but the debate still holds of under what conditions the elite genotypes have to be selected. If the conditions of plant growth act as a signal for methylation and inactivation or demethylation and reactivation of regions in plant genome then, depending on the environmental conditions, there will be a risk of rejecting plants that are genetically superior and do not express it phenotypically because of the conditions of growth (and consequently methylation) and vice versa.

References

1. A. Leonardi, C. Damerval, Y. Hebert, A. Gallais and D. De Vienne, *Theor. Appl. Gen.*, 1991, **82**, 552

2. A.S. Tsaftaris and A. Polidoros, In Proceedings of XVI Eucarpia Maize and Sorghum Conference (A. Bianchi, E. Lupotto & M. Motto), 1993, 283

3. H. Cedar, "DNA methylation. Biochemistry and biological significance", Razin A., Cedar H., Riggs A.D., Springer Verlag Inc., New York, 1984, 147

4. A.P. Papadopoulou, PhD Thesis, Aristotle University of Thessaloniki, 1995

5.   A.C. Fasoulas, "The honeycomb methodology of plant breeding", A.C. Fasoulas, Thessaloniki, 1988

6. J. Karlinsey, G. Stamatoyannopoulos and T. Enver, *Anal. Biochem.*, 1989, **180**, 303

7. F.C. Belanger and A.G. Hepburn, *J. Mol. Evol.*, 1990, **30**, 26

8. C.V. Patel and K.P. Gopinathan, *Anal. Biochem.*, 1986, **164,** 164

9. W. Doerfler, *Ann. Rev. Biochem.*, 1983, **52,** 93

10. Y. Gruenbaum, T. Naveh-Many, H. Cedar and A. Razin, *Nature,* 1981, **292**, 860

11.   R.M. Amasino, M.C. John, M. Klaas and D.N. Crowell, "Nucleic acid methylation" Clawson G.A., Willis D.B., Weissbach A. & Jones P.A., Alan R. Liss Inc., New York, 1990

The Variability of Maize Root Architecture as a Key to Differences in Root Lodging

Y. Hébert, E. Guingo, O. Argillier and Y. Barrière

STATION D'AMÉLIORATION DES PLANTES FOURRAGÈRES, INRA, F-86600 LUSIGNAN, FRANCE

1 INTRODUCTION

The susceptibility of maize plants to wind effects, in wet-soil conditions, is a trait of major concern to plant breeders. Up to 30 percent of the harvestable biomass can be left on the field after severe lodging damages. In addition, any attempts to reduce this loss can result in polluting the harvest with soil particles and thus damaging the quality of silage.

Root lodging is a consequence of deficiency of the root system at a moment when the stalk is very developed. In most of the cases, because of soil water content, the plant is uprooted, but root breaking can also occur[16]. Genetic variability is very large throughout the different earliness groups. But, only few productive and very resistant genotypes exist within early material, mainly due to the narrow genetic base. This results in a high cost of maintaining the genetic progress, which has been stimulating an intense search for new selection tools: root-pull resistance[12], root clump volume[6,11,18,19], angle of root growth from the stem[15,18], volume and number of roots[14,19]. In European conditions, counterarguments to indirect selection with these criteria have been put forward considering the poor correlations with root lodging behaviour[3,13,19]. More recently, a lateral push-test has been developed[2,4,5].

The present paper reports the work on the genetic variability of the root lodging resistance conducted at the INRA Plant Breeding Station at Lusignan (France). The genetic correlations are discussed in terms of their biological meaning and potential use in selection. The need for new and advanced investigations on related topics are also pointed out.

2 GENERAL CONSIDERATIONS

Compared with the other temperate or tropical graminaceous species, maize shows poor predisposition to good anchorage, either on nutritional or on mechanical points of view. The root system is made of thick primary roots, spreading down to a maximum of 1.5 metres below soil surface. The primary roots rapidly thin down within the thirty first centimetres in the soil, and uprooting maize plants is relatively easy.

Corroborating this description, the average biomass shoot/root ratio has been found to be approximately 75/25 (Loudet et al., unpublished data). In spite of the high susceptibility to root lodging of old maize varieties and ecotypes, some authors put forward the possibility that this situation could result from the selection pressure on aerial (mainly grain) production[17]. Very controversial results have been published about the genetic variation in photosynthesis efficiency, and it cannot clearly be stated that the improvement of plant production is largely attributable to the improvement of photosynthesis. So, the possibility arises that genetic progress in productivity could have been done through restrictions in root development.

Assuming that there are only very few genetic variations in the available root biomass, it follows that the architecture of this part of the plant could be responsible for a major amount of the variation of lodging behaviour. The anchorage ability should greatly depend on the three-dimensional architecture of primary roots: their number and size, their branching, their spatial arrangement, their mechanical resistance.

Considering that selection for root resistance has become necessary, the impact of the resulting improvement in root traits on the characteristics of the stem must be evaluated. The choice of good-rooted genetic materials could have consequences on other characteristics, such as digestibility. This point is on the way to becoming a worrying question[7].

3 THE SIGNIFICATION OF ROOT GENETIC VARIATIONS

3.1 The Parameters of the Root System

The clump of a maize plant is made of stacked internodes. Each of them bears roots at its base, and a leaf at the top (plus a bud, eventually developing an axillary stem). The shape of this part of the plant can be described by measuring the dimensions and the relative position of each of its components (Figure 1).

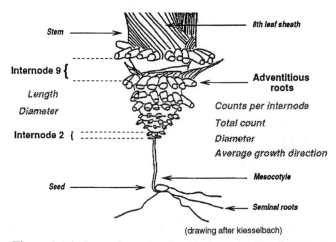

(drawing after kiesselbach)

Figure 1 *The base of a maize plant and some traits of interest.*

In our work, we usually count the roots on each internode separately (these counts add up to the total number of roots of the plant), measure the diameters of the basal sections of roots to estimate an average diameter value, and evaluate the direction of growth of the roots in order to account for the shape of the clump (either horizontal or vertical roots). These traits together are linked to the mechanical resistance of the anchorage. It should be noticed that the characteristics of root elongation and branching are not mentioned here because they are very difficult to measure in practice. Their possible role will be kept in mind in the conclusions.

One is easily aware of another difficulty in dealing with root system evaluation: all the dimension parameters (root diameter, growth direction) are affected by the internode considered. The variations from one internode to another must be taken into account. It also means that several levels should be needed for a complete evaluation of the variability.

3.2 The Genetic Variability

Genotypic differences affect all the parameters described earlier, as exampled in Figure 2. If important differences are due to environmental effects (such as location or year), very significant differences are attributable to the general combining ability of the material. In other studies, experimental hybrids as well as varieties assessed at several INRA locations also showed very large variations between them[9].

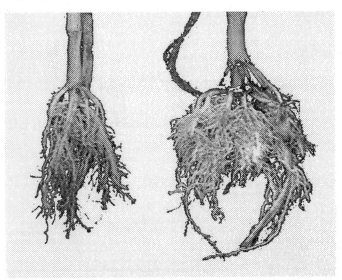

Figure 2 *Uprooted plants from different genotypes exhibit great variations.*

For some traits (root counts on upper internodes, root diameter), significant genotype by environment interaction has been noticed (even though lower than genotype and environment effects). Factorial regression analyses have been performed in order to bring to the fore co-variable traits which could be able to account for a major part of this variation[10]. In these experiments, earliness-related traits, like silking date, diameter of the stem and number of leaves, have shown responsibility for up to 90 percent of the additive genotype variation. These covariates were not so powerful in accounting for the genotype by year interaction. On the contrary, environmental covariates have proved very efficient (from 58 to 92 percent of the interaction sum of squares). Climatic conditions (precipitation, temperature) are responsible for a great part of the interaction, which proves that the climatic adaptation of the material is important to consider.

3.3 Root System and Aerial Development

One exciting question to the breeder is: "How correlated are the root traits and the aerial development traits?". Such associations, if any, could be of primary importance in breeding programmes, since they affect the breeding methods and the choice of base material.

The answer strongly depends on the genetic material in use. The wider the genetic base, the stronger the correlations. A moderate negative correlation is commonly observed between the number of roots and earliness[8]. A similar covariation applies to the diameter of the roots. But generally, the correlation coefficients are not higher than 0.5, which means that not more than a third of the root genetic variation can be attributable to earliness differences. This point is of the greatest relevance to breeding.

4 THE SELECTION FOR HIGH LODGING RESISTANCE

4.1 The Implication of Root Traits in Lodging Resistance

The correlation coefficients between root traits and lodging behaviour are generally not

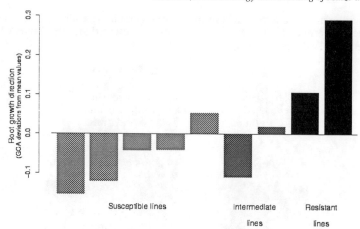

Figure 3 *Root growth direction of different inbred lines evaluated within a diallel design.*

very high[8, 9, 16]. The correlations with root counts are particularly low, probably due to the fact that what we call an "elongated root" is not always a sufficiently grown and branched root, that is, a mechanically active root.

The root growth direction seems to be a worthwhile trait to consider in order to describe the variation in root lodging (Figure 3). This trait makes it possible to distinguish between resistant and susceptible genotypes (notice that intermediate values cannot be used to classify the corresponding individuals). The average diameter of the primary roots is also an interesting parameter. Nevertheless, as shown on Figure 4, non-linear relationships could possibly account for the moderate correlation coefficients obtained in most of the cases.

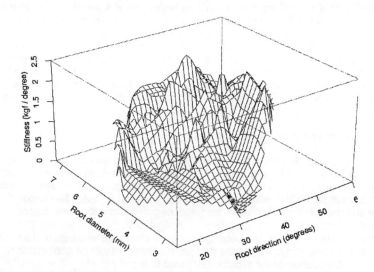

Figure 4 *Response surface of the measure of stiffness of stalk according to the growth direction and the diameter of the roots on internode 7 (20 F1 hybrids, 270 plants).*

On a statistical point of view, this situation restricts the power of linear prediction models. Thresholds can clearly be seen on Figure 4. They should be incorporated in equations in order to improve the quality of lodging prediction.

4.2 Relationships With Aerial Traits

Positive correlations have been obtained between root lodging and the length of some internodes of the stem base[8]. The interpretation of such an association is difficult in that two different explanations can be given: (*i*) the lengths of the 7th and 8th internodes are components of the root architecture, the longer the internodes, the higher the resistance moment of the roots they bear; (*ii*) the observed relationships result from the variation in earliness. It should be noticed that these relationships are not linear either.

Another interesting result of our works is the positive relationship between early vigour and actual lodging ratings[8]. The negative correlations observed could result from the selection for better early vigour previously undergone by the material involved in the experiments. Such selection could lead to (at least temporarily) a weakened root system, to the advantage of the stalk.

The relations with ear and plant heights, and with biomass are controversial. We have found very low and negative correlation coefficients, whereas Melchinger *et al.*[13] reported significant positive ones, either on inbred lines or on hybrids. The genetic base of the material under study must be put forward to account for such divergent results.

4.3 What Criteria for Indirect Selection ?

The various correlations computed make it possible to describe root lodging as a complex multi-trait phenomenon. Several morphological components are implied in the expression of the character, so that indirect selection could hardly be effective.

Even within narrow earliness ranges, traits like root growth direction and root diameter always appear satisfactorily correlated with *in situ* root lodging. In spite of the existence of thresholds, the variations of the associated traits can be used in a predictable way, since the relationships between the dependant character and the independent variables can generally be seen as successive linear models. Accordingly, indirect selection can be used to screen material and efficiently discard the worst genotypes. The question of prediction remains when material of intermediate value is considered. Owing to the correlations, the power of all the uni-variable prediction models is insufficient.

Multi-trait selection can also be done on the basis of root system characters assessed on uprooted clumps or not. This has been experienced at Lusignan, in trials and nurseries. Encouraging results have been obtained[8]. Other results on hybrids[1] do not really support the efficiency of the method: a qualitative evaluation of the development of aerial roots (with regard to their potential contribution to the quality of anchorage) enabled us to retain best lodging resistant parent lines (multi-local GCA evaluation with testers). But good genotypes were also discarded because other compensating traits were in effect but not visible. The interest of such a procedure in nursery is different (Barrière, unpublished data) since the aim is to choose the best parents for pedigree breeding. In this situation, significant genetic progress is guaranteed by the use of the best material, especially since good correlations exist between line and hybrid values[9]. For the improvement of root lodging resistance, a general evaluation of the quality of the clumps provides adequate combinations of several favourable characters. The interest of a global judgement of the anchorage is that a kind of natural equilibrium can be thus found.

Nevertheless, the general constraints of the variability must be kept in mind: the genetic progress in early material (the range of variability in which the problem of root lodging is the most worrying) will always be more expensive than in later material. As the obvious consequence of low average values, the genetic variance will always be poor.

5 CONCLUSION

It has been shown that the level of susceptibility of maize genotypes to root lodging largely depends on architectural parameters of the root system, together with characters of the aerial part. The environmental conditions also play an important role. Genotype by environment interactions are generally moderate, so genotypes can be compared throughout locations.

Only limited prediction possibilities exist. Early prediction seems hazardous because the genetic variability of root/shoot balance changes over growth period. So, it is important to know more about the phenomenon itself. It is now clear, from what we have shown in this paper, that the morphological variations, however important they could be in practice, are only a part of the variability available to maize breeder. We must now concentrate on the other possible aspects: (*i*) the simultaneous elaboration of root and shoot parts, several patterns being expected, with consequences on the quality of root anchorage; (*ii*) the mechanical characteristics of root and stem tissues, and their relations with forage quality; (*iii*) the methodological implications of indirect selection, either with root traits or indirect global tools like the push-test.

References

1. O. Argillier, PhD thesis, Institut National Agronomique Paris-Grignon, 1995.

2. A. Duparque, A. Fouéré, and S. Pellerin, In M. Borin and M. Satin, editors, *3. Congress of the European Society for Agronomy*, p. 112, Abano-Padova (ITA), 1994/09/18-22, 1994. ESA, Colmar (France).

3. R.R. Fincher, L.L. Darrah, and M.S. Zuber, *Maydica*, 1985, **30**, 383.

4. A. Fouéré, A. Duparque, and S. Pellerin, In M. Borin and M. Satin, editors, *3. Congress of the European Society for Agronomy*, p. 126, Abano-Padova (ITA), 1994/09/18-22, 1994. ESA, Colmar (France).

5. A. Fouéré, A. Duparque, and S. Pellerin, *Agron. J.*, 1995, **87**, 1020.

6. H.K. Hayes and I.J. Johnson, *J. Amer. Soc. Agron.*, 1939, **31**, 710.

7. Y. Hébert, O. Argillier, and Y. Barrière, In AGPM, editor, *Colloque maïs ensilage*, p. 293, Nantes (France), 1996/09/17-18, 1996. AGPM, Paris (France).

8. Y. Hébert, Y. Barrière, and J.C. Bertholeau, *Maydica*, 1992, **37**, 173.

9. Y. Hébert, A. Duparque, and S. Pellerin, In D. Picard, editor, *Physiologie et production du maïs*, p. 215, Pau (France), 1990/11/13-15, 1990. AGPM, INRA, Versailles (France).

10. Y. Hébert, C. Plomion, and N. Harzic, *Euphytica*, 1995, **81**, 85.

11. J.R. Holbert and B. Köhler, *J. Agr. Res.*, 1924, **27**, 71.

12. T.C. Kevern and A.R. Hallauer, *Crop Sci.*, 1983, **23**, 357.

13. A.E. Melchinger, H.H. Geiger, and G.A. Schmidt, *Maydica*, 1986, **31**, 335.

14. G.J. Musich, M.L Fairchild, V.L. Fergason, and M.S. Zuber, *Crop Sci.*, 1965, **5**, 601.

15. H.G. Nass and M.S. Zuber, *Crop Sci.*, 1971, **11**, 655.

16. S. Pellerin, R. Trendel, and A. Duparque, *Agronomie*, 1992, **10**, 439.

17. K.H.M. Siddique, R.K. Belford, and D. Tennant, *Plant and Soil*, 1990, **121**, 89.

18. D.L. Thompson, *Agron. J.*, 1968, **60**, 170.

19. M.S. Zuber, In *Corn and Sorghum Research Conference*, 1968, volume 23, p. 1.

Physiological and Environmental Considerations when Breeding for Low-input Conditions in Maize

A. Elings, G. O. Edmeades and M. Bänziger

THE INTERNATIONAL MAIZE AND WHEAT IMPROVEMENT CENTER, APDO. POSTAL 6-641, 06600 MÉXICO DF, MÉXICO

1. INTRODUCTION

Global population will increase to about 12-13 billion in the year 2100, of which about 10-11 billion will be living in developing countries. Although theoretical studies indicate that sufficient food can be produced for a population of this size, socio-economic, political, biotic and abiotic factors will limit actual productivity[2]. Grain yields of maize, which is grown extensively in the tropical and sub-tropical environments of the developing world, have been rising since 1960: from 1.5 to 3.1 t ha^{-1} in West Asia and North Africa; from 1.2 to 2.8 t ha^{-1} in South, East and Southeast Asia, and from 1.2 to 2 t ha^{-1} in Latin America. Major reasons for increased grain yields are the greater uses of fertilizer and improved germplasm. Yield increase in sub-Saharan Africa was relatively low, viz. from 0.9 to 1.2 t ha^{-1}. Reasons are the limited use of fertilizer and improved germplasm, and intercropping, which may have disguised some of the increases in land productivity[8].

Table 1 *Regional importance of abiotic growth limiting factors. Summary of discussions between 70 NARS scientists representing 32 maize growing countries held during the symposium on 'Developing Drought and Low N-Tolerant Maize', CIMMYT, March 1996. 1 = most important growth limitation.*

Growth limitation	lowland Latin America	highland Latin America	Asia	lowland Eastern Africa	mid-elevation Eastern Africa	lowland Western Africa	mid-elevation Southern Africa
drought	1	1	1	1	1	2	2
acid soils	3		3	4		3	3
low fertility	2	1	2	2	2	1	1
flooding	4		3				
high temp.				3			
salinity	4						
erosion						4	

Efficient use of natural resources is considered an essential element of sustainable agriculture. CIMMYT's Physiology and Stress Breeding Units therefore try to develop maize germplasm that uses water and nutrients, particularly nitrogen, more efficiently. The availability of water and nitrogen are major limitations to maize productivity in the developing world (Table 1). Whereas the potential grain yield level of about 10-15 t ha^{-1} is determined by aerial CO_2 concentration, incoming solar radiation, temperatures, and crop characteristics, the attainable yield level of about 2-5 t ha^{-1} results from limitations of water and nutrients, and a further decline to actual yield level of 1-1.5 t ha^{-1} is caused by the presence of pests, diseases, weeds and pollutants. It is within such limitations that breeding operates.

2. PHYSIOLOGICAL AND ENVIRONMENTAL BACKGROUNDS

2.1 Pre-flowering

Water availability is critical to crop establishment, as the shallow root systems can not capture water from deeper soil layers if rainfall ceases after emergence. Plant death at this stage will result in irregular plant density and sub-optimal yields. Establishment of roots, stems and leaves, and most nitrogen uptake from the soil occurs before flowering, the latter especially in the case of limited nitrogen and/or water availability. Development of a large canopy before flowering may lead to a great evaporative demand and water deficit after flowering. A conservative pre-flowering crop growth under dry conditions may allow crop growth after flowering. Nitrogen that has been taken up before flowering can be mobilized from vegetative to generative tissues after flowering, and is therefore a less critical factor, and water conserved prior to flowering may offset stress at this critical stage.

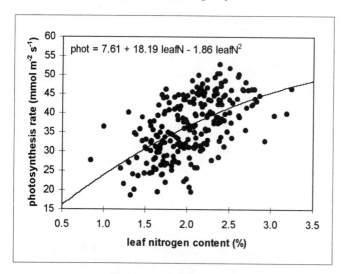

Figure 1 *The mathematical approximation between leaf nitrogen content (%) and leaf photosynthesis rate (mmol m^{-2} s^{-1}) in 18 (sub-)tropical maize genotypes. The bulked data set represents 9 open-pollinated populations, 4 hybrids, and 5 inbred lines.*

The positive effect that improved nitrogen uptake has on crop growth and production is explained through the positive relation between leaf nitrogen content and photosynthesis rate (Figure 1; see also reference 10). However, as a high leaf nitrogen content is usually accompanied by a large canopy in which the lower leaves receive little radiation and contribute little to carbohydrate production, photosynthesis rate of the entire canopy is only responsive to greater N uptake at relatively low levels of canopy nitrogen (in kg ha^{-1} surface area). The consequence for breeding is that excessive amounts of green leaf area do not add to productivity and that reduction of leaf area per plant of large crops such as maize may increase productivity if this is combined with greater plant density.

2.2 Flowering

The maize plant is very sensitive to water deficits during flowering. The developing kernels at that moment a relatively weak sink[14]. If crop growth rate around flowering declines, carbohydrate availability to the ear is reduced, which leads to two combined responses. Firstly, floret growth slows and silk exertion is delayed, though anthesis is unaffected; secondly, kernel abortion is increased[3]. This leads to a reduced number of kernels per ear (KPE) and sometimes complete barrenness, both of which contribute to reduced grain yield. Even if growing conditions improve afterwards, new kernels or ears can not be formed. Only a greater kernel weight can compensate, and there are quite conservative limits to how large kernels can grow.

Thus the anthesis-silking interval (ASI) increases under drought stress, but it is non-causally related to KPE and grain yield[5,9]. The number of ears per plant (EPP) is also highly correlated with grain yield[6]. As both characters are heritable under both poor growing conditions, and are easy to observe, they form important secondary selection criteria.

2.3 Post-flowering

Maximum grain yield, which is determined by the maximum kernel weight, EPP and KPE, is limited by the amount of carbohydrate produced in the leaves after flowering and that available for mobilization from the stem reserves. If limited nitrogen or water availability leads to accelerated leaf senescence, crop growth and grain filling rates will fall. The ratio between transpiration and potential evapotranspiration is linearly related to crop growth rate[11]. If therefore under severe drought conditions transpiration can not be maintained, crop growth and grain filling may cease entirely. We therefore breed for extended leaf longevity under low N and drought and unrolled leaves under drought. Both are considered an indication that water uptake is maintained, e.g. through a good root system or osmotic adjustment.

3. MANAGING THE ENVIRONMENT

We are making use of two experimental stations in México, located at Poza Rica with a humid lowland tropical climate (PR; 20° N, 87° W, 60 masl) and at Tlaltizapán with a mid-altitude tropical climate (TL; 18° N, 99° W, 940 masl). Drought evaluations take place at TL during the dry winter season from December to May. Three water regimes are created: 1) full irrigation that meets the evaporative demand by the crop, 2) intermediate water deficit, by withholding irrigation from 2 weeks before anthesis onwards, and 3) severe water deficit, by withholding irrigation from 3 weeks before anthesis onwards. Low nitrogen evaluations take place at PR during both the summer season from July to October and the winter season, in fields that have not received artificial nitrogen application for the past 9 years. Because a high nitrification rate in this warm and humid tropical environment leads to relatively large amounts of available nitrogen at sowing, we sow sorghum (summer) or bread wheat (winter) alongside the maize. The root systems of these crops compete with the maize root systems for nutrients, thus reducing early-season nitrogen availability for the establishing maize crop.

4. DROUGHT

4.1 Tuxpeño Sequía

Selection began in 1975 in the elite lowland tropical white dent population Tuxpeño Sequía. This population underwent 8 cycles of recurrent full-sib (FS) selection in the dry summer season at TL. Each of 250 FS families was grown under the 3 water regimes. Selection of the best 60-80 families was based on an ideotype with high grain yield, delayed foliar senescence, and constant anthesis date under all water regimes, and high leaf and stem extension rates, high rate of silk extrusion, reduced canopy temperatures, and reduced ASI under drought. These drought-related characters were combined with disease and lodging tolerance. Selection resulted in a significant increase in grain yield of 108 kg ha^{-1} cycle^{-1} at yield levels ranging from 1 to 8 t ha^{-1}. The same rate of change was observed under well-watered and drought-stressed conditions[4]. In international testing, gains averaged 90 kg ha^{-1} cycle^{-1} at a mean yield level of 5.6 t ha^{-1} [7], which indicates that 83% of the gains from selection in the dry winter season of TL is carried over to more normal summer rainy seasons in the tropics. This example shows that selection for improved grain yield under drought can be combined with maintained grain yield under well-watered conditions. This is important as it enables farmers in marginal environments to benefit from occasional high rainfall.

Table 2 *Response per selection cycle (%, p < 0.05) in 8 cycles of recurrent full-sib selection in Tuxpeño Sequía[4,5].*

Trait	Drought	Well-watered	Trait	Drought	Well-watered
Total biomass	ns	0.8	Tassel biomass at anthesis	-1.7	-2.6
Grain yield	8.9	3.4	Days to 50% anthesis	-0.5	-0.5
Harvest index	8.2	1.7	Days to 50% silking	-3.2	-0.9
Kernels per plant	9.3	1.7	Anthesis-silking interval	-16.1	-23.0
Ears per plant	9.4	0.5	Radiation use efficiency	ns	1.6
Kernels per ear	ns	1.2	Indicators of water status	ns	ns
Weight per kernel	ns	1.7	Leaf chlorophyll conc.	ns	ns
Ear biomass at anthesis	14.7	6.5			

The response per cycle of various plant characters is given in Table 2. Gains in grain yield have essentially been achieved by altering dry matter partitioning. Total above-ground dry matter production has remained constant. A greater fraction of carbohydrates is allocated earlier to the developing kernels, which has resulted in more ears per plant (reduced barrenness), more kernels per ear and greater kernel weight under well-watered conditions, and a reduced ASI under drought. Radiation use efficiency, indicators of plant water status (leaf water potential, leaf osmotic potential, soil water content), and leaf chlorophyll content did not change. In spite of attempts to maintain anthesis date, this has shown a slight advance. Early flowering and maturity imply escape from late-season drought, and therefore lead to higher grain yields. However, as early germplasm can not make use of occasional late-season rainfall, we prefer to develop germplasm that is tolerant to late-season drought, and is late enough to capture late-season rainfall and a greater amount of radiation.

4.2 Secondary traits
Evaluation of the value of secondary traits in cycles of selection of a number of populations under drought indicate that, leaf erectness and tassel size combine high heritability, with a low to moderate adaptive value; that ASI, grain yield, barrenness (number of ears per plant) and senescence combine a relatively high heritability with a moderate to high adaptive value; and that leaf osmotic concentration, canopy temperature, leaf extension rate, leaf chlorophyll concentration and leaf area combine a low to moderate heritability with a low to moderate adaptive value. It is therefore the second group of characters that we currently use in our index selection for drought tolerance.

Our selection scheme has moved from recurrent full-sib selection to recurrent S_1-selection (see Betran et al., this volume), which facilitates inbred line extraction. We have a wide range of populations under improvement, including earlies and lates, yellows and whites, flints and dents, and populations that combine tolerance to drought and low nitrogen. We have demonstrated that gains of 100 kg ha^{-1} year^{-1} can be generally expected from selection for tolerance to mid-season drought.

5. LOW NITROGEN

5.1 Across 8328 BN
The low nitrogen breeding approach and results are comparable with those of drought. Population Across 8328 BN is the low nitrogen ('Bajo Nitrogeno' in Spanish) version of Across 8328, which in 1983 was formed from a high-yielding fraction of Population 28, a late yellow tropical dent[12]. Full-sib selection of 230-270 families of Across 8328 BN was started in 1986 at PR. The selection index comprised high grain yield under high and low N; reduced ASI, increased leaf chlorophyll concentration, delayed leaf senescence under low N; and stable anthesis date, plant and ear height[13]. More recently, increased EPP under low N has been added to our selection index. Three cycles of selection resulted in 1.9% kg ha^{-1} cycle^{-1} change in biomass (averaged over high and low N situations), 2.3% kg ha^{-1} cycle^{-1} change in grain yield, -1.5% cycle^{-1} change in florets ear^{-1}, 1.6% change cycle^{-1} in kernels ear^{-1}, 0.7% g m^{-2} cycle^{-1} change in total N content, and 1.5% g m^{-2} cycle^{-1} change in grain N content. Nitrogen uptake under low N did not improve through selection, perhaps because all available soil N was utilized by the crop. The nitrogen harvest index was unchanged with selection. As under drought, more carbohydrates are partitioned to the ear as a result of selection. In addition, N mobilization from non-leaf plant parts to the larger grain sink increased[13].

5.2 Selection under low vs. high N conditions
The relative efficiency of selection under high N for gains under low N was calculated for 15 progeny trials including populations, topcrosses, lines and hybrids[1]. If grain yields are reduced by 23% (or 43%, if a 95% confidence interval is taken into account) of those under high N, then selection for yield under low N becomes more efficient in the low N environment. However, selection under low N is complicated by a reduced broad-sense heritability. Analysis shows that this decreasing heritability with reduced nitrogen

availability is not caused by increased error variance, which remains relatively stable at about 0.4, but by a reduction of genetic variance from 0.32 at high N to 0.12 at low N^1.

6. CONCLUSIONS

It is possible to select for increased grain yields under water or nitrogen limited growing conditions, and to combine this with maintained or improved grain yields under better growing conditions. To achieve this, selection must be based on an index that takes into account yields at these two production levels and a number of secondary traits with moderate to high adaptive value and heritability. Experience suggests that the use of controlled screening environments contribute considerably to efficacy of the selection process.

References

1. M. Banziger, J. Betran and H.R. Lafitte, Crop Sci., 1997, in press.
2. P. Buringh, H.D.J. and van Heemst, 'An Estimation of World Food Production Based on Labour-Oriented Agriculture', Centre for World Food Market Research, Wageningen, 1977, 46 p.
3. P. Bassetti and M.E. Westgate, Crop Sci., 1993, **33**, 279.
4. J. Bolaños and G.O. Edmeades, Field Crops Res.,1993a, **31**, 233.
5. J. Bolaños and G.O. Edmeades, Field Crops Res.,1993b, **31**, 253.
6. J. Bolaños and G.O. Edmeades, Field Crops Res.,1996, **48**, 65.
7. P.F. Byrne, J. Bolaños, G.O. Edmeades and D.L. Eaton, Crop Sci., 1995, **35**, 63.
8. CIMMYT, '1989/90 CIMMYT World Maize Facts and Trends: Realizing the Potential of Maize in Sub-Saharan Africa', CIMMYT, Mexico.
9. D.P. DuPlessis and F.J. Dijkhuis, S. Afric. J. Agric. Sci., 1967, **10**, 667.
10. H. van Keulen and N.G. Seligman, 'Simulation of Water Use, Nitrogen Nutrition and Growth of a Spring Wheat Crop', Simulation Monographs, PUDOC, Wageningen, 1987, 47.
11. H. van Keulen and J. Wolf (Eds.), 'Modelling of Agricultural Production: Weather, Soils and Crops', Simulation Monographs, PUDOC, Wageningen, 1986, 120.
12. H.R. Lafitte and G.O. Edmeades, Field Crops Res., 1994a, **39**, 1.
13. H.R. Lafitte and G.O. Edmeades, Field Crops Res., 1994b, **39**, 15.
14. R.H. Schusser and M.E. Westgate, Crop Sci., 1995, **35**, 1074.

The Role of Environmental Conditions on Quantitative and Qualitative Characteristics of Silage and Corn Maize Varieties

J. Van Waes and E. Van Bockstaele

CLO-GENT, PLANT BREEDING INSTITUTE, DEPARTMENT OF VARIETY TESTING, BURG. VAN GANSBERGHELAAN 109, B-9820 MERELBEKE-LEMBERGE, BELGIUM

1 INTRODUCTION

In the European Union, a new variety of an agricultural crop must submit to official tests for D.U.S. (Distinction, Uniformity, Stability) and V.C.U. (Value for Use and Cultivation) before commercialisation[1,2]. The performance of a variety can depend on environmental factors. In many cases there are important significant genotype by environment interactions (GxE). These interactions complicate the advice service for the farmers and make it inconvenient to find the best genotypes in breeding programmes. Therefore, the importance of stability of maize hybrids has been recognized[3].

This paper deals with the influence of environmental factors on the behaviour of silage and corn maize varieties in Belgium. Furthermore, GxE interactions were carried out to verify the stability of maize varieties. Varieties which give security (= stability) across the years are wanted by the farmers.

2 MATERIALS AND METHODS

2.1 Plant material and organization of the trials

The results are based on four years of official trials, carried out in different agricultural regions in Belgium[4,5]. Each year about 150 varieties, with a broad range for earliness, were tested. The testing period for a new maize variety is at least two years and can be prolonged to a third year[6,7]. For silage maize the total dry matter yield, earliness (% dry matter of the whole plant), resistance to lodging, digestibility and starch content were evaluated. For corn maize the grain yield, earliness (% dry matter of the grains), resistance to lodging and to Fusarium stalk rot were evaluated.

2.2 Statistical analysis of the genotype by environment interactions

The stability was carried out with Eberhardt-Russell analyses[8]. Each year-location combination is considered as an environment. For each variety a linear regression is calculated, based on the average of the variety in each environment and on the global average of all

Genetics, Biotechnology and Breeding of Maize and Sorghum

environments. Furthermore the slope of the regression line is calculated. A slope equal to one corresponds to an average stability, a value > 1 means a low and a value < 1 a high stability.

3 RESULTS

3.1 Corn maize

3.1.1Earliness (% dry matter of the grains). A lot of consistency for variety ranks was recorded when considering the percentage of the dry matter of the grains across all locations. For 1992 nearly the same variety ranks were recorded across the six locations (figure 1). The consistency for variety ranks was also very high for the three other years (1993, 1994 and 1995). Looking across the years the consistency was high for most varieties, taking into consideration that differences of three to four ranks are not significant.

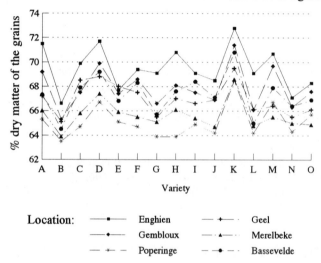

Figure 1 *Earliness of corn maize varieties as function of the location - Trials 1992*

3.1.2 Grain yield. The consistency for variety ranks of grain yield across the locations was not very good. For 1992 the lines for the six locations were not very parallel, so there are important variety by location interactions (figure 2). Especially for Hollogne, several varieties were deviant. Only half of the varieties had a stable yield across the locations. The consistency for variety ranks was also very good in 1993, 1994 and 1995. Looking across the years the consistency for variety ranks was good for half of the tested varieties.

3.1.3 Resistance to lodging and to Fusarium stalk rot. Evaluation of resistance to lodging is not easy but it is an important agricultural factor. In addition to the fact that lodging happens occasionally, the intensity and its geographical distribution are not predictable[9]. Variety classification for this characteristic is not very valuable after one year of observation because of large differences between locations and years. To know the resistance for lodging, an evaluation across at least three years is necessary. As for lodging, the intensity and geographical distribution for resistance to Fusarium stalk rot are unpredictable. The percentage of attacked plants can vary strongly from one location to another. Across the three years variety ranks with clearly significant differences can be detected. The consistency across the years is in general better in contrast with resistance to lodging.

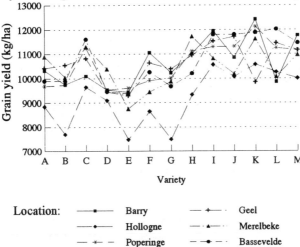

Figure 2 *Grain yield of corn maize varieties as function of the location - Trials 1992*

3.2 Silage maize

3.2.1 Earliness (% dry matter of the total plant). As for grain maize a lot of consistency for variety ranks was found when considering the percentage of the dry matter of the whole plant across the locations. For 1992 nearly the same variety ranks were recorded across the seven locations (figure 3). The intersections between the lines for the different locations were rare. The consistency was also good in the three other years. Looking across the years the consistency for the percentage of the total dry matter was good for most varieties.

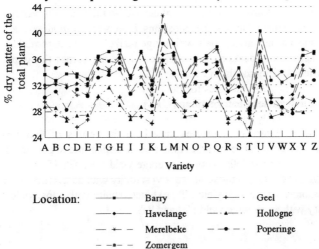

Figure 3 *Earliness of silage maize varieties as function of the location - Trials 1992*

3.2.2 Total dry matter yield. The consistency across the locations was in general not very good. For 1992 many intersections between the lines for seven locations were recorded (figure 4). For 1994 we have seen the same pattern. The consistency for variety ranks was somewhat better in 1993 and in 1995. Looking across the years the consistency for variety ranks was in most cases good.

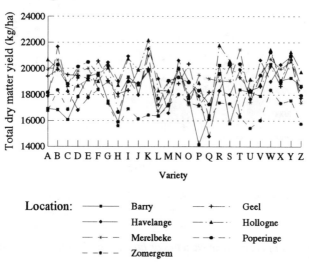

Figure 4   *Total dry matter yield of silage maize varieties as function of the location - Trials 1992*

3.2.3 Resistance to lodging. As for grain maize, variety classification for this characteristic with silage maize is only valuable with data of at least three years. Looking across the years, the variety ranking was in most cases consistent.

3.2.4 Digestibility and starch content. Looking across the years the variety ranks were consistent. The consistency across the locations is in general moderate as worked out in a study for optimizing the variety trials with silage maize[9]. As for digestibility, the consistency for starch across the locations is moderate[10]. Looking across the years the variety ranks for most varieties were consistent.

3.3 Genotype by environment interactions (GxE)

3.3.1 Corn maize. After GxE analyses, based on the results of four years with six locations per year, the corn maize varieties could be classified into six groups for the grain yield (figure 5) : a) average stability with an average yield : 'F', 'G', 'H' and 'L' ; b) average stability with a high yield : 'D' ; c) moderate to low stability with an average yield : 'B', 'J', 'K' and 'M' ; d) low stability with a high yield : 'I' ; e) high stability with a low yield : 'A', 'C' and 'O'; f) high stability with an average yield : 'E' and 'N'.

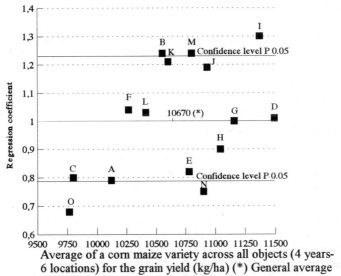

Average of a corn maize variety across all objects (4 years-
6 locations) for the grain yield (kg/ha) (*) General average

Figure 5 *Plot-regression coefficient for GxE analyses vs. variety (average across all objects, 15 corn maize varieties)*

3.3.2 Silage maize. After GxE analyses, based on the results of three years with seven locations per year, the silage maize varieties could be classified into seven groups for the total dry matter yield (figure 6) : a) average stability with an average yield : 'G', 'P', 'J', 'X', 'M', 'U', 'Q', 'K' and 'H' ; b) average stability with a high yield : 'A', 'F', 'S' and 'W' ; c) moderate to low stability with an average yield : 'D', 'N' and 'R' ; d) low stability with a high yield: 'B' ; e) high stability with an average yield : 'C' and 'L' ; f) moderate to high stability with an average yield : 'E' and 'V' ; g) average stability with a low yield : 'I', 'O' and 'T'.

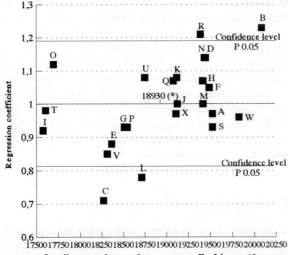

Average of a silage maize variety across all objects (3 years-7 locations) for the total dry matter yield (kg/ha) (*) General average

Figure 6 *Plot-regression coefficient for GxE analyses vs. variety (average across all objects, 24 silage maize varieties)*

4 DISCUSSION

Agronomists base their recommendations in most cases on regional yield means or from field means using data from the nearest crop testing location to the farmer[11]. Therefore the importance of stable performance of maize hybrids across different locations is important[3] and it is valuable to include stability in selection criteria.

There is a worldwide interest among plant breeders, geneticists and production agronomists in genotype by environment (GxE) interactions. A stability analysis is often conducted to estimate and interpret GxE interaction.

We have analysed the most important agronomical characteristics of corn and silage maize varieties to determine genotype by environment (location, year) interactions. For corn maize the consistency of variety ranks for earliness was in general good across the locations and the years. Variety ranking for grain yield was more influenced by the location and in a lower content by the year. Because of important genotype by environment interactions it is necessary to have a sufficient number of official trials each year and not to reduce this, so as to predict in a reliable way the agronomical value of a new variety[10]. Classification for resistance to lodging and to Fusarium stalk rot was only valuable when considering data of at least three years. Especially, the intensity of lodging is unpredictable. It is however a very important characteristic for the security of a maize culture[9].

Variety ranking for earliness with silage maize was also very consistent across the locations and the years. The total dry matter yield was more influenced by the location. However, the differences were not so great as for grain yield. Stress conditions are probably not so harmful for silage as for corn maize. A lower cob yield can be compensated by a higher stem and leaf yield[10]. For resistance to lodging, the same remarks as for corn maize are applicable to silage maize. The variety rankings for the qualitative characteristics digestibility and starch content, are for most varieties consistent over the years. It is however important to analyse only samples of very regular trials (low variation coefficient for the total dry matter yield) in order to evaluate real genotypical differences for quality and not influences of stress conditions.

For corn and silage maize the percentage of stable varieties was much greater in the second trial year in comparison with the first year. This is probably due to the specific criteria for admission which have an effect on the selection for admission of varieties with a moderate to good stability. The GxE analyses led us to conclude that for corn maize most of the varieties have a moderate to good stability. It is remarkable that the most versus least stable varieties have, in general, the lowest versus highest potential yield. For silage maize we have recorded that the majority of the tested varieties have a good stability with an average yield. The number of varieties with a very good or bad stability is limited. The criteria for admission have an important influence on the type of admitted varieties ; in general, varieties with a good stability score the best.

5 CONCLUSION

Environmental conditions have, in general, an important influence on the most important agronomical characteristics of silage and corn maize. The genotype by environment interactions complicate the agronomical service. For the security of the culture, stable varieties are of interest. Security means stability under different climatological conditions, with a guaranteed earliness, a stable yield (especially cob yield) and a good resistance for lodging and for diseases and a good stress resistance (f.e. against drought).

With the genotype by environment analyses we can give to the farmers objective information about the stability of new varieties. It is advisable to follow the evolution in the variety assortment. Top varieties of maize stay for maximum three to four years at the top, so that little by little a switch-over to the best new varieties is necessary so as to profit continuously from the progress in plant breeding.

Acknowledgements

The authors wish to thank Mrs Nancy De Vooght for her technical assistance and Mrs Martine Van Bogaert for typing the manuscript and the general lay-out.

References

1. Ano. Guidelines, UPOV, 1969, 25 pp.
2. Ano. E.C. directives, 1972, 35 pp.
3. T. Hallauer. Proc. XVth Euc. Congress, Maize and Sorghum, 1990, 31-71.
4. J. Van Waes and A. De Vliegher. Reports Corn maize, RvP - CLO-Gent, 1992-1995, 45 pp.
5. J. Van Waes and A. De Vliegher. Reports Silage maize, RvP - CLO-Gent, 1992-1995, 84 pp.
6. J. Van Waes, A. De Vliegher and E. Van Bockstaele. Descriptive and recommended list Corn maize. RvP - CLO-Gent, 1992-1995, 9 pp.
7. J. Van Waes, A. De Vliegher, E. Van Bockstaele and L. Carlier. Descriptive and recommended list Silage maize. RvP - CLO-Gent, 1992-1995, 17 pp.
8. S.A. Eberhardt and W.A. Russel. Crop Science, 6, 1966, 36-40.
9. J. Van Waes. PhD thesis CLO-RvP, 1995, 440 pp.
10. J.L. Gellner, F.A. Chalick and J.J. Bonnenman. Plant Var. and Seeds, 1991, 4 : 67-72.
11. J. Van Waes. Proc. Symp. Coop. de Pau, 1992, 227-241.

SESSION V
BREEDING MAIZE FOR STRESS TOLERANCE
AND LOW INPUT DEMANDS

Breeding Maize for Stress Tolerance and Low Input Demands

E. Lupotto, C. Balconi, M. Maddaloni and M. Motto

ISTITUTO SPERIMENTALE PER LA CEREALICOLTURA, SEZIONE DI BERGAMO, 24126
BERGAMO, ITALY

1 INTRODUCTION

Breeders have made spectacular progress in improving the performance of maize hybrids
during the last six decades. More than 50% of the increase in yield has been attributed to
genetically improved cultivars; the remainder derives from greater and more efficient use of
chemical fertilizers and crop management. Evidence has indicated that hybrid selection
based on conventional breeding procedures was effective in improving both plant vigor and
adaptation to varying environmental conditions [1,2,3].
   Environmental and biological stresses have a great negative impact on maize
production throughout the world. In addition, the growing concern about the environment,
together with a strong indication to lower production costs, encourages the development of
cultivar crops, that require few chemicals. Breeding strategy, selection criteria and efficient
methods that would allow maximum progress in a short time are critically important for
developing a maize plant more tolerant to biotic and abiotic stresses and with low input
demands. In particular, recent advances in genetics, cell and molecular biology provide new
opportunities to enhance adaptation of maize plants to environmental stresses and in
developing more efficient plants. It is expected that new plant technology will have a
significant impact on maize improvement by supplementing the present activities of
breeders, in expanding and diversifying the gene pool of maize, by introducing specific
genes and shortening the time required for the production of new varieties and hybrids
capable to cope with adverse environmental conditions. In this paper we will illustrate and
discuss with selected examples the creation of new maize varieties with a better adaptability
to disease and pest damages, water shortage, high or low temperatures, and with low input
demands.

2 BIOTIC STRESSES

2.1 Insect Resistance

A maize crop is exposed to a complex array of insects from the time it is planted until it
is utilized as food or feed. A review of the important insect species attacking maize has been
given by DICKIE and GUTHRIE[4]. Improvement in resistance and/or tolerance of maize
cultivars to some insect species by breeding has been reasonably successful. Probably the
most significative success has been the development of maize cultivars that have resistance
to the European corn borer (ECB, *Ostrinia nubilalis,* Hubner). ECB is a major pest of
maize in North America and Europe, and yield loss of 3 to 7% per borer per plant can result
from ECB feeding at various stages of plant growth[5].
   Natural host plant resistance factors to ECB exist, but they provide a general inadequate
control. Leaf feeding resistance to first generation ECB depends mainly on antibiosis of

first and second instar larvae; in some inbreds it was associated with high concentration of the aglucone DIMBOA (cyclic hydroxamate, 2,4-dihydroxy-7-methoxybenzoxozin-3-one)[6]. Resistance to the second generation ECB is associated with reduced sheath collar feeding and stalk tunneling, but the underlying mechanisms are largely unknown[7]. Extensive research has permitted the development of several resistant inbred lines and breeding populations that have been released for breeding purposes[8,9,10,11]. Although these cultivars have not contributed directly to hybrid improvement they can be used as breeding sources of resistance.

While many hybrids with good resistance to first-generation leaf feeding damage have been developed, most current hybrids are still susceptible to sheath and collar feeding of later-generation borers. The corn synthetic BS9(CB)C4 was selected specifically for resistance to ECB throughout the life cycle of the plant, and has been an important germplasm source in host-plant resistance investigations [10,12]. Quantitative resistance factors for later-generation feeding damage have been identified in the inbred lines B52 and De811 through molecular mapping techniques, and efforts are underway to transfer these resistance loci into elite germplasm [13]. Progress has also been made in developing hybrids which will stand well and retain ears despite heavy infestations of late season borers. Stay-green may also reflect improvement for resistance to second-generation ECB. DUVICK[2,14] showed a continuous improvement in the resistance to second-generation ECB from the 1930 to the 1980 hybrids, which would be conducive to greater stalk quality and reduced harvest loss.

Genes encoding insecticidal proteins have been isolated from *Bacillus thuringiensis* (*Bt*), a Gram positive, spore forming soil microbe[15]. These genes, when modified for expression in plants, control target pests and confer insect tolerance[16]. Control of ECB in maize through expression of genes of the CryIA class from *Bt* has been recently reported[17,18]. They introduced a synthetic gene encoding a truncated version of the CryIA(b) protein derived from *Bacillus thuringiensis* var. Kurstaky HD-1[19] into maize cells. The transgenic plants produced high levels of insecticidal protein and exhibited excellent protection against extremely high, repeated infestations with ECB. This is a new and powerful tool which is already used in integrated pest management strategies directed against the insect attack. As agricultural biotechnology continues to mature, new strategies for insect control will become available.

2.2 Disease Resistance

Most of the diseases of maize are caused by viruses, bacteria, and fungi. SMITH and WHITE[20] have given a comprehensive review of diseases occurring in maize. Selection for disease resistance has been an integral component of maize breeding for many years, yet there are few data reflecting directly on success for this selection relative to its effect on grain yield. It was estimated that yearly losses caused by maize diseases in the United States are 7 - 17% [21]. Losses would be higher in value if breeders did not support the parental inbred lines with continuous improvements for resistance to diseases. Stay-green, which is primarily a rating for total plant health, has been evaluated in several experiments. Improvement in greater stalk strength has been well documented[14,22] and stalk quality is dependent upon the overall plant health. All experiments that have evaluated the trait "stay-green" [3,14], have shown significant improvement for overall plant health from the early to the most recent decades.

At the molecular level, a considerable body of evidence indicates that plants respond to microbial attack by elaboration of a number of active defenses, including synthesis of antibiotics, reinforcement of cell walls, and stimulation of lytic enzymes[23]. These defenses can also be activated by elicitors derived from microbial cell walls and culture fluids[23,24]. Until recently, however, no plant resistance gene had been cloned and characterized at the molecular level. Since then, resistance genes from several plant species have been cloned[25]; this constitues a major advance for molecular plant biology and may lead to the development of novel methods for disease control.

Maize leaf spot, caused by *Cochliobolus* (formerly *Helminthosporium*) *carbonum*, is emerging as one of the better understood fungal diseases of plants. The interaction between

C. carbonum race 1 and maize is genetically the best characterized gene-to-gene interaction described for maize[26]. The specificity of this disease is determined by the cyclic tetrapeptide HC-toxin produced by the fungus. Fungal isolates producing the toxin are exceptionally virulent on plants homozygous recessive at the *Hml* locus. Maize that is heterozygous or homozygous dominant at *Hml* is 100 time less sensitive to the toxin[27]. JOHAL and BRIGGS[28] generated and cloned several *Hml* alleles by transposon-induced mutagenesis. The wild-type gene shares homology with dehydroflavonol-4-reductase genes of petunia, and snapdragon. The results indicate that *Hml* encodes a reduced nicotinamide phosphate (NADPH)-dependent HC-toxin reductase, HTRC. HTRC inactivates HC-toxin, produced by the fungus *C. carbonum* Nelson race 1 that permits the fungus to infect certain genotypes of maize[28,29]. The discovery that *Hml* encodes a reductase that inactivates HC-toxin establishes a mode of action of plant disease resistance genes that may be typical of other natural resistance genes. Genes that encode resistance to specific disease may be engineered in the laboratory, as it has been done for virulence factors.

In maize, twenty-five different genetic specificities have been identified that determine gene-for-gene resistance to *Puccinia sorghi*, the common leaf rust pathogen of maize. These specificities have been mapped to five separate locations in the maize genome, indicating the existence of at least five rust resistance genes that have been named: *Rp1, Rp 3, Rp4, Rp5,* and *Rp6* [30,31]. *Rp1, Rp5,* and *Rp6* (plus a related gene, *Rpp9,* that specifies resistance to *Puccinia polysora* [32] are found clustered within a region of approximately 5 cM near the end of the short arm of maize chromosome 10[30,31]. *Rp3* and *Rp4* have been mapped near the centromere of chromosome 3 and on the long arm of chromosome 4, respectively [31]. The availability of DNA markers already mapped to areas tightly linked to *Rp1* and *Rp3,* has allowed a genetic analysis of the fine structure of these loci[33,34]. These studies have demonstrated that *Rp1* is a complex locus, probably composed of multiple *Rp* genes (*Rp1-A* to *Rp1-N*) and that the meiotic instability of this locus[35,36,37,38] is associated with unequal crossing over between tandemly repeated DNA sequences in the *Rp1* region[34], and the appearance of new rust resistance specificities seems to be associated with recombination in the *Rp1* complex [39]. By contrast, *Rp3* appears to be a relatively simple and stable locus[40]. Further results[37] indicate that the hypersensitive response elicited by *Puccinia sorghi* requires a cell autonomous, non diffusable factor specified by the *Rp1* locus. To isolate such resistance genes, various approaches are being pursued, including map-based cloning and transposon tagging[41].

A gene encoding for a pathogenesis-related protein was cloned from a cDNA library enriched in aleurone-specific sequences prepared from maize seeds two days after germination[42]. The accumulation of its mRNA increases after rehydration of desiccated seeds. Moreover, a close relationship exists between the expression of this gene and attack of the fungus *Fusarium moniliforme*, a common pathogen of maize that causes stalk and ear rot[43]. The increased expression of the gene could be part of a defense mechanism against pathogens active during seed germination.

Plant inhibitors of gene translation, called ribosome-inactivating proteins (RIPs), are active against fungal and viral growth. Plant RIPs inhibit protein synthesis in target cells by a specific modification of 28S rRNA. As a consequence, elongation factor 2 binds less efficiently and the elongation step in protein synthesis is inhibited[44]. By sequence analysis and *in vitro* assay several workers have reported that the maize b-32 protein, a 32-kDa endospermic albumin under the control of the gene *O2*, is a ribosome-inactivating protein[45,46,47]. This protein may play a role against pathogenesis, as evident from the increased susceptibility of *O2* mutant kernels (and thus deficient of RIP) to fungal attack[48] and insect feeding[49]. Transgenic tobacco plants expressing the b-32 protein show an increased tolerance to the infection of the soil-born fungal pathogen *Rhizoctonia solani*[50]. Research to manipulate this gene for protecting maize plants is in progress in our laboratory.

Undoubtedly, increasing knowledge of the molecular basis for both pathogenicity and resistance will help to produce novel forms of disease resistance.

2.3 Virus Resistance

Plant viruses have a negative impact throughout the world on crop production. Maize dwarf mosaic virus (MDMV) is the most widely distributed and important virus disease of maize affects both growth and yield of maize varieties[51]. Control or reduction of MDMV can be accomplished with the use of resistant maize cultivars, eliminating overwintering hosts, early planting to avoid exposure to insect vectors, and controlling insect vectors themselves. Genetic resistance in maize to MDMV has been demonstrated to be dominant, and from two to five genes appear to be involved in resistance to MDMV[20].

Transgenic plants expressing coat protein (*cp*) genes present a new tool for creating virus resistant plants. MURRY *et al.*[52] have reported the first example of *cp*-mediated protection in transformed maize using the *cp* gene of the MDMV strain B (MDMV-B). In this study the *MDMV-Bcp* gene was cloned into a monocot expression cassette and introduced into maize cell lines suspension culture. Fertile plants were regenerated and challenged with a virus inoculum concentration adjusted to produce symptoms in non transgenic controls. The presence of the *MDMV-cp* provided resistance to inoculations with MDMV-A or MDMV-B and to mixed inoculations of MDMV and maize chlorotic mottle virus.

Other emerging strategies with potential for control of virus infections include the use of small catalytic RNA (ribozymes), virus-mediated activation of toxic genes in transgenic host plants, over-expression of viral movement proteins. The expression of combinatorial antibodies to virus specific proteins is currently being explored in several laboratories for protection of plants from virus infections. This knowledge can expand our repertoire of methods available to protect plants against virus diseases.

3 ABIOTIC STRESSES

Despite continuing efforts to improve cultivation practices, maize plants will always be subjected to a variety of environmental extremes. In even the most productive agricultural regions, drought and temperature stresses can occur throughout the growing season, resulting in injury and reduced plant yield.

3.1 Thermal Tolerance

Exposure of plants to elevated temperatures or other environmental stresses such as cold, nutritional starvation, heavy metal exposure, induced the synthesis of a limitated number of heat shock protein(s) (hsps)[53]. Although no specific function(s) has yet been established for these proteins, it is widely assumed that hsps confer at least a transient protection against heat-induced damage[54].

In maize there are at least 10 new polypeptides synthesized at heat shock temperatures as revealed in one-dimensional gel electrophoresis. These can be further resolved in 20 to 30 spots by two-dimensional gel electrophoresis[55]. The 18 kDa HSPs of maize comprise the most prominent HSP class[56]; this class is composed of a number of isoelectric variants [57] which are considered to be encoded by a multigene family[58].

Similarly, among the high molecular weight HSPs, the cytosolic HSP70s family is encoded by multiple genes which are closely related[59]. Studies of the hsp response in maize have addressed the aspects of induction temperature, tissue specificities, hsp profile and intracellular localization[55,56,57,60]. It has been also found that genetic variability in hsp synthesis exists among maize genotypes[61,62]. The identification of intraspecific variation in hsp synthesis will facilitate genetic analysis of the role of hsp genes in plant thermal tolerance.

Cold tolerance of maize, i.e. the ability of a genotype to germinate and grow vigorously after emergence under cool spring conditions, is recognized as an important plant property affecting final grain yield[64]. Information has revealed that sizeable genetic differences exist among maize strains in germination rate, speed of emergence, and seedling vigour, suggesting that it is possible to improve these traits[65].

The physiological and biochemical responses of maize to low temperatures during germination and seedling growth have been extensively investigated[65]. These studies indicate that numerous processes such as root growth, nutrient uptake, supply of phytohormones, cell division and extension, and photosynthetic activity are negatively affected, to various extents, by low temperatures; however, the biochemical nature of cold-tolerance is complex and not fully understood. PRASAD *et al.*[67], investigating the molecular mechanisms of chilling acclimation in maize seedlings, isolated cDNA representing three chilling acclimation-responsive (CAR) genes. Identification of one of the CAR genes, as catalase (*Cat 3*) led them to hypothesize that chilling may be imposing oxidative stress in the seedling. CHRISTIE *et al.*[68] have considered the activation of general phenylpropanoid and anthocyanin pathways in relation to low temperature stress but further research is needed to clarify these relationships.

3.2 Water Deficit

In maize, the effects of moisture stress on vegetative growth and grain production have frequently been described[69]. Collectively, these studies have revealed that the magnitude of the effect depends both upon the level of stress and on the stage of plant growth.

Exploitation of genetic differences in traits contributing to drought tolerance may be a means of improving maize crops for drought-prone environments. Reports have shown the existence of genotypic variability in several parameters used as drought tolerance indicators[70,71,72]. Some of the traits studied and discussions of their role in drought tolerance and possible application to breeding programs include: leaf conductance and osmoregulation, leaf size and rolling, leaf water potential, stomatal sensibility, root growth and soil moisture extraction, heat and desiccation tolerance, abscisic acid and amino acids[73,74]. Although some of these specific traits may be associated with drought response, their relationship with grain yield is not sufficient to merit major screening programs.

Leaf epicuticular wax (EW) is a trait related to drought resistance. Resistance to diffusion of water through the cuticle of some species is completely determined by the wax content of the cuticle[75]. Cuticle water loss becomes important in the maintenance of high leaf water potential, although water loss via the cuticle is usually small compared with that lost through the stomata, even during drought when stomata are closed. Drought resistance may be increased if cuticle resistance to transpiration during drought can be augmented by higher EW content[76]. Drought resistance has been shown to be a function of EW content in sorghum[77]. However, the biosynthesis of wax formation is poorly understood, as well as the function of epicuticular leaf waxes in preventing water losses: mutants in epicuticular wax biosynthesis are termed "*glossy*", and the term "*glossy genes*" is used to indicate the genes responsible of the specific phenotype. For a deeper understanding of the role of the *glossy genes*, these are being actively investigated, and tagged by transposons in our laboratory for their physical isolation. TACKE *et al*[78] starting from the allele *gl2-m2*, have been recently successful in cloning *Gl2*. The predicted glossy-2 protein consists of 426 amino acids. No similar amino acid sequence was found in protein data banks and the biochemical function of the *Gl2* gene product is still unknown.

Differences in the expression of specific genes between stress-sensitive and stress-tolerant plants indicate that tolerance is conferred by genetically encoded mechanisms[88]. In particular, these studies indicated that some genes expressed during stress promote cellular tolerance to dehydratation activating protective functions in the cytoplasms, or with alteration of cellular water potential to promote water uptake, control of ion accumulation, and further regulation of gene expression[79]. In maize, genes encoding dehydration-related proteins referred to as dehydrins have been cloned and characterized[80,81]. These proteins, which may be structure stabilizers with detergent and chaperon-like properties[82], are characterized by the consensus KIKEKLPG amino acid sequence found near the carboxy terminus, and usually repeated from one to many times within the protein. Many dehydrins, including the maize RAB-17, contain a tract of serine residues, and it is at these positions where phosphorylation has been shown[83].

The consideration of the effect of water stress on the maize seed has been also followed using a special class of maize mutants known as viviparous kernels[84] in which ABA biosynthesis is altered. Results support a clear role of ABA regulation in dehydrin expression[85,86].

The availability of *viviparous (vp)* mutants has opened new horizons to study regulatory factors active during water deficit in gene activation. The *Vp1* gene encodes a transcriptional activator required for ABA induction of maturation-specific genes leading to the acquisition of desiccation tolerance[87], and it is actively studied for its significance in ABA-mediated drought tolerance in plants[88,89].

Although these studies are promising, it continues to be difficult to ascertain the actual function of drought-induced gene products. A considerable research effort is still needed to define the role of these proteins in the acquisition of drought tolerance, including their structural and enzymatic functions. Also the mechanisms regulating the expression of these genes in response to different environmental stimuli should be better understood. However, the molecular approach to these problems has already revealed information which permits the monitoring of the level of plant tolerance to cold, heat, drought and salts. In other plant species, studies have modeled the overexpression of specific proteins, and have analyzed their potential for reducing the effects of water stress using physiological measurements in transgenic plants[90]. Such studies have revealed two promising strategies for increasing water-stress tolerance: 1) engineering increased content of osmolytes, and 2) overexpressing enzymes to increase oxygen-radical scavenging in chloroplasts.

3.3 Oxidative Stress

A frequently observed consequence of environmental adversity is the phenomenon of oxidative stress. By perturbing cellular metabolism and in particular photosynthetic processes, many stress factors induce the production of a reactive oxygen complex and include both enzymic and non-enzymic components (reviewed in[91]). The ability of plants to detoxify radicals under conditions of water deficit or other oxidative stresses, is probably the most critical requirement.

Superoxydismutase (SOD), catalases, and various peroxidases such as guaicol peroxydase and ascorbate peroxydases constitute the cellular defense against oxidative stress. SODs catalyze the first step in active oxygen-scavenging systems. In maize, SOD isozymes are coded for by nonallelic nuclear genes[92,93,94]. The available information suggests that any efforts to engineer more oxitolerant plants should perhaps focus less on over- or under-expressing SOD levels, and more on maintaining the endogenous optimal levels of all SOD isozymes when the plant is under oxidative stress, and in keeping a well-balanced and coordinated expression of all essential antiodant enzymes in the various cell compartments. Molecular analyses of *SOD* gene structure and expression may help in understanding the mechanisms regulating the expression of the individual *SOD* genes at the appropriate location and time in the plant.

4 UPTAKE AND EFFICIENCY OF UTILIZATION OF NUTRIENTS FROM THE SOIL

A major strategy in the attainment of present high yield of maize hybrids has been the selection of strains which demonstrate favorable responses to high N fertilizer applications[95]. However, the costs of N fertilizers and concern for environmental hazards associated with excessive fertilization make it desirable to find alternative ways to lessen the dependence on high N inputs for obtaining maximum yields. One of the alternatives available is to breed maize plants which can absorb and use soil N more efficiently.

The pattern of accumulation and redistribution of N in maize plants have been studied by many investigators[96]. These results have demonstrated that although maize can absorb substantial quantities of N following anthesis, mobilization of vegetative N accumulated before anthesis provides the major source of N in the grain. Maize plants are, in fact, able to accumulate large amounts of N in stalks, leaves, and roots during the vegetative growth phase[97]. During grain-filling N undergoes export, in various degrees, to the developing seeds ranging from 60 to 85% of the total N present at anthesis. It has been observed that N in the stalk of maize during vegetative growth can constitute as much as 40-50% of total-N content of the plant[98]. Clearly, at least in some circumstances, the stalk is an important temporary reservoir for N that can be remobilized for ear development. TA[99] by studying N metabolism in the stalk tissue of maize after anthesis, has found that glutamine synthase (GS), specifically the GS1 isoform, plays a central role in the remobilization of nitrogenous compounds from the stalk to the ear.

Current work indicates that the efficiency with which maize plants utilize N fertilizer is affected by several parameters including root morphology and extension, biochemical or physiological mechanisms regulating NO_3 uptake, redistribution and transport of N to different plant parts[100,101]. Therefore, the potential for developing superior N-efficient hybrids seems to exist. In addition, there are clear indications that genotypic differences in N utilization exist in maize as is demonstrated by differential responses to N fertilizer and by differences in absorpion, accumulation, and in utilization of absorbed N[100,102,103,104]. These genotypic differences suggest the possibility for developing superior N-efficient hybrids.

Recently, several genes encoding enzymatic activities of the pathway of N metabolism have been identified and cloned in maize. These include the structural gene for nitrite reductase[105], nitrate reductase[106], glutamine synthase[107] and glutamate synthase[108]. Although several key elements remain also to be identified at the molecular level, for instance nitrate and amino acid transport processes, it may be useful to modify by genetic engineering the flow of metabolic pathway involved in N metabolism. This approach may be pursued either by overexpression or by changing tissue specificity of expression or by reducing the expression of one of the cloned N-related genes, playing a central role in N metabolism. This may allow the study, in a coordinated effort, of the physiological consequences of these modifications on N utilization efficiency, plant growth and yields. From this analysis it should be possible to derive rational approaches leading to the production of more efficient crops for future agricultural needs.

5 PROSPECTS

Maize breeders have been extremely effective in increasing yield during the last six decades. Studies suggest that past yield improvement in maize can be attributed, to a large extent, to increased stress tolerance and that the potential for yield improvement with further increases in stress tolerance is good. However, further analysis of stress tolerance is required to reach more definite conclusions regarding the role of stress tolerance in the genetic improvement of maize. The genetic basis of stress tolerance will likely be of great interest in the future to plant breeders, molecular biologists and plant physiologists.

The application and extension of biotechnology to crop protection has already led to the field testing of genetically modified maize hybrids which express the Bt inseticidal protein, conferring resistance to ECB. These new techniques hold great promise for crop protection, and increased resistance or tolerance to biotic and abiotic factors are anticipated through the use of biotechnology. These techniques have already been successfully applied in several areas and in the future will certainly supplement and complement conventional breeding in improving tolerance of maize cultivars to environmental stresses and in developing maize plants with low input requirements.

Acknowledgements. Research cited from author's laboratory was financially supported from grants of Ministero delle Risorse Agricole, Alimentari e Forestali, Roma, special grant: Programma Nazionale "Resistenze genetiche delle piante agrarie agli stress biotici ed abiotici".

References

1. T. M. Crosbie, Proc. 37th Ann. Corn and Sorghum Industry Res. Conf. Chicago, IL H. D. Loden and D. Wilkinson (Eds), 5-9 Dec. Am. Seed Trade Assoc. Washington, DC, 1982, p. 206.
2. D. N. Duvick 'Genetic contributions to yield gains of five major crop plants', W. R. Fehr (Ed.) Spec. Publ. No 7, Crop. Sci. Soc. Am. Madison, Winsconsin, USA, 1984, p. 15.
3. W. A. Russell, *Maydica*, 1985, **30**, 85.
4. F. F. Dickie and W. D. Guthrie, 'Corn and corn improvement', G.F. Sprague and J.W. Dudley (Eds), Am.Soc.Agron.Madison, Winsconsin, 1988, 3rd Ed., p. 767.
5. R. E. Lynch, *J. Econ. Entomol.* 1980, **73**, 159.
6. J. A. Klun and J. R. Robinson, *J. Econ. Entomol.*, 1969, **62**, 214.
7. G. R. Pesho, F. F. Dickie and W.A. Russell, *Iowa State J. Sci.*, 1965, **40(1)**, 85.
8. W. A. Russell, *Crop Sci.*, 1976, **16**, 315.
9. W. A. Russell and W. D. Guthrie, *Crop Sci.*, 1979, **19**, 565.
10. W. A. Russell and W. D. Guthrie, *Crop Sci.*, 1982, **22**, 694.
11. W. A. Russell and W. D. Guthrie, *Crop Sci.*, 1991, **31**, 238.
12. J. R. Klenke, W. A. Russell and W.D. Guthrie, *Crop Sci.*, 1986, **26,** 864.
13. C.C. Schoen, M. Lee, A. E. Melchinger, W. D. Guthrie, and W. L. Woodman, *Hered.*, 1993, **70**, 648.
14. D. N. Duvick, *Maydica*, 1992, **37**, 69.
15. H. Hofte and H. R. Whiteley, *Microbiol. Rev.*, 1989, **53**, 242.
16. F. Salamini and M. Motto, 'Plant Breeding. Principles and prospects', M. D. Hayward, N. O. Bosemark, I. Romagosa (Eds), Chapman & Hall, London, New York, 1993, p. 138.
17. M. G. Koziel, G, L. Beland, C. Bowman, et al., *Bio/Technology*, 1993, **11**, 194.
18. C. Armstrong, G. B. Parker, J. C. Pershing, et al., *Crop Sci.*, 1995, **35**, 550.
19. M. Geiser, S. Schweitzer and C. Grimm, *Gene*, 1986, **48**, 109.
20. D. R. Smith and D. G. White, 'Corn and corn improvement', G. F. Sprague and J. W. Dudley (Eds), Am. Soc. Agron. Madison, Winsconsin, 1988, 3rd Ed., p. 687.
21. M. C. Shurtleff, 'Compendium of Corn Diseases', Am. Phytopathol. Soc. St Paul, Minnesota, USA, 1980, 2nd Ed.
22. D. C. Tapper, PhD Thesis, Iowa State University of Ames, 1983.
23. R. A. Dixon and C. J. Lamb, *Ann. Rev. Plant Physiol. Plant Mol. Biol.,* 1990, **41**, 331.
24. C. J. Lamb, M. A. Lawton, M. Dron and R.A. Dixon, *Cell*, 1989, **56**, 215.
25. B. J. Stackawicz, F. M. Ausubel, B. J. Baker, J. G. Ellis and J. D. G. Jones, *Science,* 1995, **268**, 661.
26. O. E. Nelson and A. J. Ullstrup, *J. Heredity,* 1965, **55**, 195.
27. R. P. Scheffer, R. R. Nelson and A. J. Ullstrup, *Phytopathol.*, 1967, **57**, 1288.
28. G. S. Johal and D. P. Briggs, *Science*, 1992, **258**, 985.
29. R. B. Meeley, G. S. Johal, S. P. Briggs, and J. D. Walton, *Plant Cell*, 1992, **4**, 71.
30. B. H. Lee, A. L. Hooker, W. A. Russell, J. G. Dickinson and A. L. Flangas, *Crop Sci.,* 1963, **3**, 24.
31. D. R. Wilkinson and A. L. Hooker, *Phytopathol.*, 1968, **58**, 605.
32. A. J. Ullstrup, *Phytopathol.*, 1965, **56**, 425.
33. S. H. Hulbert and J. L. Bennetzen, *Mol. Gen. Genet*, 1991, **226**, 377.
34. M. A. Sudupak, J. L. Bennetzen and S. H. Hulbert, *Genetics*, 1993, **133**, 119.
35. A. Pryor, *Trends Genet.*, 1987, **3**, 157.
36. J. L. Bennetzen, W. E. Blevins and A. H. Ellingboe, *Science*, 1988, **241,** 208.
37. J. L. Bennetzen, M. M. Qin, S. Ingels and A.H. Ellingboe, *Nature*, 1988, **332**, 369.
38. K. S. Hong, T. E. Richter, J. L. Bennetzen and S. H. Hulbert, *Mol. Gen. Genet.*, 1993, **239**, 115.
39. T. E. Richter, T. J. Pryor, J. L. Bennetzen and S.H. Hulbert, *Genetics*, 1995, **141**, 373.
40. K. M. S. Saxema and A. L. Hooker, *J. Genet. Cytol.*, 1974, **16**, 857.

41. J. L. Bennetzen and J. D. G. Jones, 'Genetic Engineering, Principles and Methods', J. K. Setlow (Ed), Plenum Press, New York, 1992, Vol. 4, p 99.
42. J. M. Casacubarta, P. Puigdomenech and B. Sansegundo, *Plant Mol. Biol.*, 1991, **16**, 527.
43. T. A. Kuchareck and T. Kommedhal, *Phytopathol.*, 1966, **56**, 983.
44. F. Stirpe, L. Barbieri, M. G. Battelli, M. Soria and D. A. Lappi, *Bio/Technology*, 1992, **10**, 405.
45. M. Maddaloni, L. Barbieri, S. Lohmer, M. Motto, F. Salamini and R. Thompson, *J. Genet. & Breed.*, 1991, **45**, 377.
46. H. W. Bass, C. Webster, G. R. O'Brian, J. K. M. Roberts and R. S. Boston, *Plant Cell*, 1992, **4**, 225.
47. T. A. Walsh, A. E. Morgan and T.D. Hey, *J. Biol. Chem.*, 1991, **266**, 23422.
48. J. R. Loesch, P. J. Foley and D. F. Cox, *Crop Sci.*, 1976, **16**, 841.
49. S. C. Gupta, V. L. Asnani and B. P. Khare, *J. Stored Prod. Res.*, 1970, **6**, 191.
50. M. Maddaloni, F. Forlani, V. Balmas, G. Donini, L. Stasse, L. Corazza and M. Motto, *Plant Cell Rep.*, 1996 (submitted).
51. J. K. Knocke, R. E. Gingery and R. Louie, 'Plant Diseases of international importance. Diseases of cereals and Pulses', A. N. Mukhopadhyay, J. Kumar and H. S. Chaube (Eds), Prentice Hall, Englewood Cliffs, N.J., 1992, p. 235.
52. L. E. Murry, L. G. Elliott, S. A. Capitant, et al., *Bio/Technology*, 1993, **11**, 1559.
53. M. M. Sachs and T-H. D. Ho, *Ann. Rev. Plant Physiol.*, 1986, **37**, 363.
54. J. K. Key, J. A. Kimpel, C. Y. Lin, R. T. Nagao, E. Vierling, E. Czarnecka, W. B. Gurley, J. K. Roberts, M. A. Mansfield and L. Edelman, 'Molecular Biology of Plant Stress', J. K. Key and T. Kosuge (Eds), Alan R. Liss, N. Y., 1985, p. 161.
55. P. Cooper, T-D. H. Ho and R. M. Hauptmann, *Plant Physiol.*, 1984, **75**, 431.
56. B G. Atkinson, L. Liu, I. S. Goping and D. B. Walden, *Genome,* 1989, **31**, 699.
57. C. L. Baszczynski, D. B. Walden and B. G. Atkinson, *Can. J. Biochem.*, 1982, **60**, 569.
58. I. S. Goping, J. F. Frappier and D. B. Walden, *Plant Mol. Biol.,* 1991, **16**, 699.
59. E. E. M. Bates, P. Vergne and C. Dumas, *Plant Mol. Biol.*, 1994, **25**, 909.
60. P. Cooper and T-H. D. Ho, *Plant Physiol.*, 1983, **71**, 215.
61. J. A. Jorgensen, J. Weng, T-H. D. Ho and H. T. Nguyen, *Plant Cell Rep.*, 1992, **11**, 576.
62. C. Frova, G. Taramino, G. Binelli and E. Ottaviano, *Maydica*, 1988, **33**, 65.
63. H. T. Nguyen, M. Krishnan, J. J. Burke, D. R. Porter and R.A. Vierling, 'Environmental stress in plants', J. H. Cherry (Ed), Springer Verlag Berlin, Heidelberg, 1989, p. 319.
64. J. J. Mock and M. McNeill, *Crop Sci.*, 1979, **19**, 239.
65. G. L. Pozzi, E. Gentinetta and M. Motto, *Plant Breed.*, 1986, **97**, 2.
66. P. Stamp, 'Advances in Agronomy and Crop Science', G. Geiger (Ed), Paul Parey, Berlin, 1984, Vol. 7.
67. T. K. Prasad, M. C. Anderson, B. A. Martin and C. R. Steward, *Plant Cell*, 1994, **6**, 65.
68. P. J. Christie, M. R. Alfinito and V. Walbot, *Planta*, 1994, **194**, 541.
69. E. W. Daw, M Sc. Thesis, Univ of Guelph, Ontario, Canada, 1981.
70. T. V. Williams, R. S. Snell and J. F. Ellis, *Crop Sci.*, 1967, **7**, 179.
71. T. V. Williams, R. S. Snell and J. F. Ellis, *Crop Sci.*, 1969, **9**, 19.
72. A. J. Hall, C. Chimenti, N. Trapani, F. Vilella and R. Cohen de Hunau, *Field Crops Res.,* 1984, **9**, 41.
73. R. A. Fisher, and N. C. Turner, *Ann. Rev. Plant Physiol.*, 1978, **29**, 277.
74. R. A. Fisher and A. F. E. Palmer, 'Symp. on Potential productivity of field crops under different environments', IRRI, Philippines, 1978.
75. J. Schoenherr, *Planta*, 1976, **131**, 159.
76. J. A. Clark and L. Levitt, *Physiol. Plant.*, 1956, **9**, 598.
77. A. Blum, 'Plant Breeding for stress environments', CRC Press, Boca Raton, Florida USA, 1988.

78. E. Tacke, C. Korfhage, D. Michel, M. Maddaloni, M. Motto, S. Lanzini, F. Salamini and H.P. Döring, *Plant J.,* 1995, **8**, 907.
79. E.A. Bray, *Plant Physiol.,* 1993, **103**, 1035.
80. T. J. Close, A. A. Kortt and P. M. Chandler, *Plant Mol. Biol.,* 1989, **13**, 95.
81. R. Asghar, R. D. Fenton, D. A. Demason and T. J. Close, *Protoplasma,* 1994, **177**, 87.
82. T. J. Close, *Physiol. Plant.,* 1996, **97**, 795.
83. J. Vilardell, A. Goday, M. A. Freire, M. Torrent, M. C. Martinez, J. M. Torné and M. Pages, *Plant Mol. Biol.,* 1990, **14**, 423.
84. D. S. Robertson, *Genetics,* 1955, **40**, 745.
85. R. Paiva and A.L. Kriz, *Planta,* 1994, **192**, 332.
86. Z. Y. Mao, R. Paiva, A. L. Kriz, J. A. Juvik, *Plant Physiol. Biochem.,* 1995, **33**, 649.
87. D. R. McCarty, T. Hattori, C.B. Carson, V. Vasil, M. Lazar and I.K. Vasil, *Cell,* 1991, **66**, 895.
88. U. Hoecker, I. K. Vasil and D. R. McCarty, *Genes & Develop.,* 1995, **9**, 2459.
89. V. Vasil, W.R. Marcotte, L. Rosenkrans, S.M. Cocciolone, IK. Vasil, R.S. Quatrano and D.R. McCarty, *Plant Cell,* 1995, **7**, 1511.
90. H. J. Bonhert and R. G. Jensen, *Trends in Biotech.,* 1996, **14**, 89.
91. J. G. Scandalios, *Plant Physiol.,* 1993, **101**, 7.
92. D. Zhu and J. D. Scandalios, *Free Rad. Biol. Medicine,* 1995, **18**, 179.
93. D. Zhu and J. D. Scandalios, *Plant Physiol.,* 1994, **106**, 173.
94. L. Guan and J. D. Scandalios, *Plant J.,* 1993, **3**, 527.
95. R.H. Hageman,'Nitrogen assimilation of plants', E. J. Hewitt and C.V. Cutting (Eds), Academic Press, New York, 1979, p. 591.
96. M. Motto and F. Salamini, *Rivista di Agronomia,* 1981, **15**, 3.
97. J. W. Friedrich and L. E. Schrader, *Agron. J.,* 1979, **71**, 466.
98. W. L. Pan, J. J. Camberaro, W. A. Jackson and R. H. Moll, *Plant Physiol.,* 1986, **82**, 247.
99. C. T. Ta and R.T. Weiland, *Crop Sci.,* 1991, **32**, 443.
100. W. A. Jackson, W. L. Pan, R. H. Moll and E. J. Kamprath, 'Biochemical basis of plant breeding', C. A. Neyra (Ed), CRC Press Boca Raton, Florida, 1986, p. 73.
101. C.A. Neyra, P. A. Pereira and J. I. Baldani, 'Maize breeding and maize production', Euromaize '88, Belgrade, Yugoslavia, 1988, p. 219.
102. R. H. Moll and E. J. Kamprath, *Agron. J.,* **69**, 81.
103. W. G. Pollmer, D. Eberhard, D. Klein and B. S. Dhillon, *Crop Sci.,* 1979, **19**, 82.
104. A. G. Reed and R. H. Hageman,, *Plant Physiol.,* 1980, **66**, 1179.
105. K. Lahners, V. Kramer, E. Back, L. Privalle and S. Rothstein, *Plant Physiol.,* 1988, **88**, 741.
106. G. Gowri and W. H. Campbell, *Plant Physiol.,* 1989, **90**, 792.
107. M-G. Li, R. Villemur, P.J. Hussey, C. D. Silflow, J. S. Gantt and D. P. Snustad, *Plant Mol. Biol.,* 1993, **23**, 401.
108. H. Sakakibara, M. Watanabe, T. Haset and T. Sugiyama, *J. Biol. Chem.,* 1991, **266**, 2028.

Combined Selection Based on HS, S1 and TC Family Evaluation Under Low Nitrogen Input Conditions in Maize

C. K. Goulas,[1] D. Deliporanidou,[1] N. Katsadonis,[2] J. Sfakianakis,[2] H. Karamaligas,[3] N. Katranis,[3] E. Mpletsos[2] and A. C. Gertsis[4]

[1] UNIVERSITY OF THESSALY, SCHOOL OF PRODUCTION SCIENCES, FACULTY OF AGRICULTURE, CROP AND ANIMAL PRODUCTION, GR 383 33 VOLOS, GREECE
[2] NATIONAL AGRICULTURAL RESEARCH FOUNDATION (NAGREF), CEREALS INSTITUTE, THESSALONIKI, GREECE
[3] NATIONAL AGRICULTURAL RESEARCH FOUNDATION, PALAMAS, GREECE
[4] TECHNOLOGICAL EDUCATION INSTITUTE (TEI), SCHOOL OF AGRICULTURAL TECHNOLOGY, THESSALONIKI, GREECE

1. INTRODUCTION

Maize breeders face the challenge to improve an array of agronomically important crop traits including yield, plant type, maturity and resistance to biotic and abiotic stresses. Maize hybrids having low nitrogen requirements, while maintaining agronomically acceptable yield performance under nitrogen stress and non stress conditions, are of considerable interest because of economic and environmental concerns.

Agronomic studies on hybrid response to supplemental nitrogen varied considerably, were inconclusive and the nitrogen rate by hybrid interaction was not established [1,7]. In spite of this evidence, it was presented that supplemental nitrogen up to 330 kg ha^{-1} during the course of line development resulted in developing superior inbreds than those developed under zero nitrogen conditions [11]. In fact, hybrids having the parents as inbreds developed under high supplemental nitrogen rates had consistently better performance under all supplemental nitrogen rate treatments [12]. According to previous considerations, an efficient breeding program for developing such hybrids would require the appropriate methodology to select and develop the germplasm needed. Selection for yield component traits like nitrate reductase, harvesting index, etc. did not seem to be effective for improving performance under low N conditions [3]. On the contrary, selection for yield under nitrogen deficiency stress combined with selection for component traits and followed by testing under nitrogen stress and non stress conditions was suggested as a practical integrated approach to develop the appropriate germplasm [3]. Selection based on S2 top cross performance was suggested as promising breeding methodology to develop populations tolerant to low nitrogen stress conditions [13].

Population improvement schemes based on per se progeny evaluation in the form of S1, half sib or full sib families could be employed but, certain practical limitations (family seed quantities for replicated trials in more than one location) would restrict testing breeding values of individual plants chosen as parents for the next cycle only under

under low nitrogen input conditions. On the contrary, population improvement schemes based on combined two or three family type evaluation could overcome this disadvantage providing testing of S0 genotypes under more than one nitrogen input condition. Selection based on combined half sib and S1 evaluation was effective in improving population performance[4,8,9] although it was not successful in improving population test cross performance with unrelated tester [4]. A population improvement scheme based on combined evaluation of half sib, S1 and test cross family performance could be a more effective breeding procedure for developing germplasm with high yielding performance under nitrogen stress and non stress conditions and at the same time improving its test cross performance.

The objective of this study was to study population breeding methodology based on combined three family type evaluation as a means of developing genetic material having low nitrogen input requirements and/or better nitrogen use efficiency.

2. MATERIALS AND METHODS

Population GR-OP-35, originating from free intecrossing 15 commercial hybrids followed by four mass selection cycles for yield, was used. In 1993 approximately 2000 plants were planted in the nursery under optimum nitrogen input conditions and ear shoots were covered on two eared plants. The upper ear was self pollinated to provide S1 seed and at the same time pollen was crossed on to inbred line B73 to provide test cross (TC) seed. After two days the plant was detasseled to avoid subsequent selfing and the lower ear was open pollinated to provide half sib (HS) seed. Thus each S0 plant was represented by a set of three progeny types (S1, HS and TC). In 1994 three separate yield trials were planted, one for HS, one for S1 and one for TC family evaluation. Each trial was conducted in a different location, each one representing not only a different environment (north, north-central and central-south part of the country) but a different nitrogen fertility level as well. Testing fields were cropped with maize the two previous cropping seasons without any supplemental nitrogen to secure homogeneous soil nitrate distribution. The residual NO_3-N measured prior to planting was 106.83 and 59 kg ha^{-1}, whereas the organic matter was 1.9, 1.4 and 1.1 % for S1, HS and TC evaluation fields, respectively. Supplemental nitrogen 50 kg ha^{-1} was added to S1 evaluation field. Thus each S0 genotype was evaluated under high, medium and low nitrogen conditions. An incomplete block experimental design with 16 blocks, each with two replications, was used. Ten families were randomly assigned in each block. The same S1 plant progenies in a block in the S1 trial were kept together in a separate block in the HS and TC trials. A 17th block with four replications, including ten commercial hybrids, was also included in each testing site to provide a common estimate of each environment's yielding potential. Planting density was 65000 plants ha^{-1} and two row plots (8.0 m^2) were used. Standard cultural practices were followed throughout growing season. Plots were hand harvested, plot yields were adjusted to 15.5 % moisture level and converted to yield ha^{-1}. Data were analysed on a per block basis and the individual ANOVA pooled over blocks for each trial.

Selection was based on S1, HS and TC performance within blocks. Only those S0 genotypes which yielded above their respective block means in the three progeny evaluation (S1, HS and TC) were selected. Although the target portion selected was two genotypes out of ten in each block, meaning a total of 32 ones, the above criterion resulted in 12 genotypes selected with combined S1-HS-TC evaluation as compared with 15, 19 and 21 selected based on HS-TC, HS- S1 and S1-TC combined evaluation. S1 remnant seed was used to advance the population to C1 cycle of selection. Estimates of genetic parameters were obtained using the expected values from ANOVA. Then, family yields in each trial were corrected for block effect and adjusted for location effect using the ten hybrid average yield in each location. Using these adjusted values the yield inbreeding depression was estimated as [1- (S1/HS)] per cent, and yield heterosis as [(TC/S1)-1] per cent along with the simple yield phenotypic correlation coefficients between HS-S1, HS-TC and S1-TC. Response to selection was determined by direct comparison of C0 vs. C1 under three nitrogen regimes (N0, N120 and N240) corresponding to 0, 120 and 240 kg ha^{-1}. A split plot experimental design was used with nitrogen regimes as main plots and entries as subplots in RCB arrangement with four replications. Trial was conducted in two locations in 1996 and all testing and cultural practices were those followed during family evaluation in 1994.

3. RESULTS AND DISCUSSION

Average hybrid checks on yield performance indicated a yield loss of 5-10 % as compared to yield obtained when the available soil nitrogen is not limited (Table 1). This figure was within the range of losses estimated by leaf greenness. Hybrids had an average leaf greenness of 59 SPAD units as compared to a reported range 55-60 units which corresponded to 5% yield loss [2]. Thus our data provided evidence that yield evaluation was conducted under lower than optimum nitrogen input conditions.

Population per se yielding potential estimated from half sib progeny performance was 32% inferior as compared to average hybrid checks performance, whereas the average test cross progeny performance was practically equal to hybrids (Table 1). Average S1 progeny performance indicated a 12.7% yield inbreeding depression. Population expressed less than hybrids leaf greenness and higher ASI, whereas it was more prolific. Pertinent mean squares pooled over blocks for yield and leaf greenness along with genetic parameter estimates are shown in Table 2. Variances associated with differences among all progeny types were highly significant and indicated that genetic variance was available for both yield and leaf greenness. Thus the population used fulfilled the requirements, high yield performance, genetic variance and combining ability to justify further breeding work. Genetic variances and heritability estimates for yield based on S1 progeny evaluation were in the range of magnitude previously reported [4, 8, 10] whereas the corresponding estimates based on half sib progeny evaluation were larger than those reported from similar studies [4, 10]. From all genetic estimates, the ones calculated for the S1 progenies seem to be of the greater value. Corresponding genetic parameters for leaf greenness provided evidence that this trait is a quantitative one and could be proved interesting and useful in breeding work.

Combined selection based on the three progeny type evaluation (HS-S1-TC) resulted, as expected, in higher selection pressure as compared to the corresponding based on one progeny type evaluation (7.5% vs 20.0%) and the same holds true for the other combined selection schemes. This difference reflected the lower than previously correlations reported [5, 9] between progeny types which were r=0.19, -0.08 and 0.13 for

HS-S1, HS-TC and S1-TC respectively. In spite of this, the number of genotypes (families) selected in common by the combined schemes and each progeny per se evaluation was satisfactory. In fact, the combined selection scheme based on HS-S1-TC evaluation seemed to be the most balanced and effective having half or more of the selected genotypes in common with each one of the progeny per se evaluation. This means that from the 12 genotypes selected by the combined scheme six, eight and seven would be selected using HS, S1 or TC per se evaluation respectively.

Basing selection upon superior performance in the three progeny type evaluation trials this resulted, obviously, in lower selection differential within each trial. Despite this, the expected gain from selection showed that the three progeny combined scheme (HS-S1-TC) and the corresponding two progeny combined one (S1-TC) seemed to be equally effective as compared to HS and TC per se selection in improving both population mean and its test cross performance (Table 3). Preliminary evidence from direct comparison between C0 and C1 populations in one location showed that a genetic gain of 9.5% was realised for the combined selection based on HS-S1-TC, under zero nitrogen input conditions. The C1 population did not respond to 120 and 240 kg ha[-1] supplemental nitrogen. Data abovementioned provided evidence that the combined selection based on the three progeny evaluation was effective in improving population for low nitrogen input requirements, but more data are needed for this evidence to be confirmed.

Population average yield inbreeding depression was 12.8 % which means a 0.5 % yield decrease per 1.0 % increase in inbreeding coefficient. Our estimate was lower than previously reported values ranging [4,10], indicating that the population used could be useful for developing germplasm tolerant to inbreeding depression[14]. Comparing the different selection schemes it seemed that selection based on half sib progeny per se evaluation resulted in selected genotypes expressing high inbreeding depression accompanied by high heterotic response, whereas test cross showed genotypes expressing low inbreeding depression and high heterotic response. On the contrary, the S1 per se evaluation selected genotypes expressed inbreeding vigor and low heterosis. The genotypes selected using the combined three progeny (HS-S1-TC) evaluation expressed low inbreeding depression and heterosis as compared with those selected using the combined two progeny (S1-TC) evaluation which expressed inbreeding vigor and the same heterosis. Data abovementioned provided some evidence on the type of gene action each selection procedure is capitalising.

Concerning the possibility to use leaf greenness as an indirect selection criterion for increasing yield, our data did not provide evidence that this could be effective because of low correlation between the two traits. In spite of this, genetic variance was existing for the trait and heritability; values were high indicating that this trait could be effectively improved through S1 progeny selection. It seems that leaf greenness could be helpful for effective selection to improve yield under low nitrogen input conditions.

In conclusion, the combined selection procedure based on three progeny evaluation seemed to capitalise on all types of gene action and could be effective as intra and inter population improvement (test cross performance) breeding scheme. Selection under low nitrogen input conditions seems to be effective. Furthermore, the genotype by environment interaction seems to be controlled and thus development of germplasm with high yielding potential under nitrogen stress and non stress conditions seems to be effective. Finally, the scheme studied is directly applicable to hybrid development programs since selection based on either half sib, S1 and test cross per se evaluation or

Table 1. Mean grain yield and correlated trait performance for half sib, S1 and test cross progenies in the GR-OP-35 C0 maize composite population.

Progenies	Grain yield	Moisture at harvest	*Traits* Leaf greenness	Ears per plant	ASI
	-----t ha⁻¹-----	----g kg⁻¹----	SPAD units	------No-----	----Days----
Half sib	8.55±0.80	166±33	56±3.5	1.07±0.03	4.0±1,3
	(4.01-11.03)	(148-208)	(44-75)	(1.00-1.23)	(0.0-8.0)
S1	7.46±0.86	165±60	53±4.6	1.39±0.09	3.4±0.7
	4.26-12.04	(130-201)	(45-76)	(0.92-2.53)	(0.0-6.0)
Test cross	12.36±0.48	166±75	56±2.7	1.27±0.09	3.0±0.7
	(10.91-14.17)	(141-196)	(49-68)	(1.03-1.75)	(0.0-4.8)
Hybrids	12.54±0.46	168±56	59±1.5	1.02±0.04	3.0±0.5
Checks	(10.47-14.61)	(155-193)	(44-70)	(0.98-1.05)	(1.6-6.0)

Table 2. Pertinent mean squares from the analyses of variance along with genetic parameter estimates from half sib, S1 and test cross progeny evaluation in GR-OP-35 C0 maize composite population.

Source	df	*Yield* HS	SI	TC	*Leaf greenness* HS	SI	TC
			t ha⁻¹			SPAD units	
Families	144	3.08**	3.31**	1.54**	28.18*	137.3**	16.88
Error	144	1.28	0. 45	0.55	21.75	34.47	13.44
CV (%)		13.2	9.9	5.5	4.1	9.5	10.4
σ^2 p		1.54±0.20	1.66±0.19	0.78±0.10	14.1±2.2	68.7±8.4	8.4±1.3
σ^2 f		0.90±0.20	1.43±0.19	0.50±0.10	3.2±2.1	51.4±8.3	2.3±1.2
σ^2 A		3.6 ±0.80	5.72± 0.76		12.84 ±8.2	205.7±33.2	9.2 ±5.04
GCV		11.1	17.6	5.2	3.2	14.3	0.25
h²		0.58	0.86	0.64	0.23	0.75	0.25
UCL		0.68	0.90	0.73	0.41	0.81	0.39
LCL		0.46	0.82	0.53	0.01	0.67	0.04

*, ** Significant at the P=0.01 and P=0.05 probability level.
σ^2 p = phenotypic variance (σ^2e/r+ σ^2p), σ^2 f= genetic family variance.
σ^2 A= additive genetic variance by assuming that terms for dominance were equal to zero.
h² = heritability (σ^2 G/ σ^2p) on a family mean basis, UCL=upper 90% confidence limits.
LCL=lower 90% confidence limit, GCV=genetic coefficient of variation.

Table **3**. Average inbreeding effect and heterosis response for yield observed in parent genotypes selected following various evaluation procedures along with expected genetic gain from selection.

Evaluation	Population Performance			Inbreeding Effect		Heterosis
Procedure	HS	S1	TC	ID	IV	HET
Combined	------------	%	------------	----------------%--------------		
HS- S1-TC	12.4		6.6	6.6	-	49.9
HS-S1	10.5		5.9	8.2	-	45.1
HS-TC	5.7		5.9	12.4	-	65.9
S1-TC	12.0		6.5	-	5.7	50.0
Per se					-	
HS	8.7			23.2	-	71.7
S1		21.0		-	14.0	33.8
TC			5.4	5.4	-	75.5

combined on two, three progeny evaluation (HS-S1, S1-TC, HS-TC and HS-S1-TC) could be applied at the same time.

Acknowledgements

The financial support of the Greek General Secretariat for Research and Development in implementing this research project is greatly appreciated.

References

1. W. L. Albus and J. T. Moraghan, *J. Prod. Agric.*, 1995, **8**, 581.
2. T. M. Blackmer, J. S Schepers and G. E. Varvel, *Agron. J.*, 1974, **86**, 893.
3. P. Y. Bramel-Cox, T. Barker, F. Zavala-Garcia and J. D. Eastin. 'Plant Breeding and Sustainable Agriculture: Considerations for Objectives and Methods', Crop Science Society of America and American Society of Agronomy, Madison, 1991.
4. J. G. Coors, Crop Sci., 1988, **28**, 891.
5. L. A. Duclos and P. L. Crane, *Crop Sci.*, 1976, **8**, 191.
6. N. O. Fountain and A. R. Hallauer, *Crop Sci.*, 1996, **36**, 26.
7. C. A. Gardner et al., *J. Prod. Agric.*, 1990, **3**, 39.
8. C. K. Goulas and J. H. Lonnquist, *Crop Sci.*, 1976, **16**, 461.
9. C. K. Goulas and J. H. Lonnquist, *Crop Sci.*, 1977, **17**, 754.
10. A. R. Hallauer, *Maydica*, 1990, **35**, 1.
11. T. L. Krone and R. J. Lambert, *Maydica*, 1995, **40**, 211.
12. T. L. Krone and R. J. Lambert, *Maydica*, 1995, **40**, 217.
13. H. A. Laffite and G.O. Edwards, *Maydica*, 1995, **40**, 259.
14. S. K. Vasal, S. Srinivasan and V. Vergara, *Crop Sci.*, 1995, **35**, 1233.

Breeding for Drought Tolerance in Tropical Maize

F. J. Betrán, D. Beck, M. Bänziger, J. M. Ribaut and G. O. Edmeades

INTERNATIONAL MAIZE AND WHEAT IMPROVEMENT CENTER (CIMMYT), APDO. POSTAL 6-641, 06600 MÉXICO DF, MÉXICO

1. INTRODUCTION

Food in the tropics is facing higher demand as a consequence of rapid population growth (80 million people/year) and a declining availability of water resources and other inputs[1]. Maize grain losses due to drought in the tropics may reach 24 million tons per year, equivalent to 17% of well-watered production[2].

CIMMYT's maize program objective is to increase the productivity and sustainability of maize farming systems with particular attention to environments with limiting supply of water and nutrients. This paper deals with CIMMYT's efforts and methodologies in developing lines, hybrids and populations which are tolerant to drought (DT) at flowering and grainfilling.

2. POPULATION IMPROVEMENT

Improvement in drought tolerance at flowering in Tuxpeño Sequía (tropical white dent population) has been accomplished by full-sib (FS) recurrent selection using managed stress environments[3,4,5]. The value of individual secondary traits such as ears per plant, anthesis-silking interval (ASI) and senescence has been demonstrated by linear phenotypic correlations and divergent selection[6,7]. Encouraged by the results with Tuxpeño Sequía, selection for DT was initiated either in other elite CIMMYT maize populations or in new populations formed by recombination of several DT sources (Table 1). A recurrent S_1 selection scheme has been used where 500-600 S_1 families are screened under drought and heat in Obregón, México (27°30' N, 30 m elevation) during the summer. The 200 superior families are evaluated in Tlaltizapán, México (18°N, 940 m elevation) during winter under three water regimes: severe stress (SS), intermediate stress (IS) and well-watered conditions (WW)[5]. The best 50 S_1 families are recombined to form the next cycle of selection and advanced simultaneously to S_2 to start line extraction in an integrated manner (Figure 1). Each controlled-irrigation environment exposes different genetic variation for specific traits: the Obregón environment for upper leaf firing, tassel blasting, barrenness and husk cover; the SS environment for barrenness and anthesis-silking interval (ASI)[7]; the IS environment for leaf senescence; and the WW environment for grain yield potential. Selection is based on an index integrating different traits in different conditions, while trying to keep flowering date constant in order to avoid escape by earliness. Approximate weightings for each trait appear in Edmeades et al.[8]

Table 1 *CIMMYT's populations improved for drought tolerance and their development, original germplasm, breeding scheme, cycles of selection and characteristics*

POPULATIONS	DEVELOPMENT	ORIGINAL GERMPLASM	BREEDING SCHEME	CYCLES OF SELECTION[a]	CHARACTERISITICS[b]
TUXPEÑO SEQUÍA	Recurrent Selection of elite populations	TUXPEÑO-1 C_{12}	FS	8	LT, L, W, D
TS6	"	TUXPEÑO SEQUÍA C_6	S1	3 (+6)	LT, L, W, D
LA POSTA SEQUÍA	"	Pop. 43 C_6	S1	5	LT, L, W, D
POOL 26 SEQUÍA	"	POOL 26 C_{12}	S1	3	LT, L/I, Y, D
POOL 18 SEQUÍA	"	POOL 18 C_{15}	S1	4	LT, E, Y, D
POOL 16 SEQUÍA	"	POOL 16 C_{12}	S1 /S2	3 (+2)	LT, E, W, D
DTP1	Evaluation and recombination	BEST OF SEVERAL SOURCES	S1	3 (+4)	ST/LT, I, W/Y, F/D
DTP2	Evaluation and Introgression in DTP1	DTP1 + BEST SOURCES	S1	1 (+4)	ST/LT, I, W/Y, F/D

[a] Cycles in parentheses refer to selection by half- or full-sib (FS) methods.
[b] Adaptation (LT.- tropical, ST.- subtropical), Maturity (L.- late, I.- intermediate, E.- early), Color (W.- white, Y.- yellow), Texture (F.- flint, D.- dent).

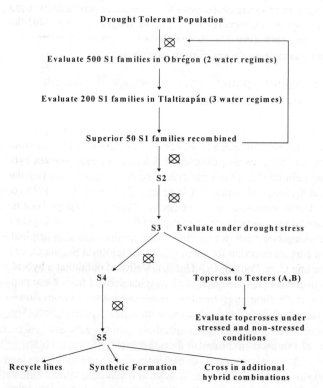

Figure 1 *General breeding scheme for drought tolerance population improvement and line development used at CIMMYT*

Recurrent selection was effective at increasing yield under both well watered and stressed environments in all populations studied (Table 2). Selection for drought tolerance under managed stressed environments has improved performance across environments incorporating broad adaptation and drought tolerance. Gains were due to a significant reduction in barrenness, an increase in grains per ear and an increase in harvest index, and were accompanied by a reduction in ASI and a delay in leaf senescence.

Table 2 *Observed genetic gains from recurrent selection under drought and well-watered conditions for several lowland tropical maize populations*[9]

		Gain cycle^{-1}: kg ha^{-1} (%)			
EVALUATIONS		**Drought**	**Well-Watered**	**Across**	**Grain yield range (t/ha)**
(1) Tuxpeño Sequía (C$_0$-C$_8$)	Managed env.	30 (8.9)	130 (3.4)	108	1.0-8.0
	Across[*]	52	101	90	0.3-7.9
(2) Late Maturing Populations:					1.0-10.4
	Tuxpeño Sequía (C$_0$-C$_8$)	80 (3.8)	38 (0.5)	59 (1.2)	
	La Posta Sequía (C$_0$-C$_3$)	229 (12.4)	53 (0.6)	141 (2.8)	
	Pool 26Sequía (C$_0$-C$_3$)	288 (12.7)	177 (2.3)	233 (4.8)	
(3) Early Maturing Populations:					1.1-6.3
	Pool 18 Sequía (C$_0$-C$_3$)	146 (10.3)	126 (2.4)	134 (3.8)	
	Pool 16 Sequía (C$_0$-C$_1$)	204 (19.2)	198 (3.6)	201 (5.8)	

[*] Gain observed across a range of environmental conditions[4].

3. DEVELOPMENT OF DROUGHT TOLERANT HYBRIDS

3.1. Line source germplasm

Edmeades et al.[10] showed that population improvement for drought tolerance in open-pollinated varieties increases the probability of obtaining stress-tolerant hybrids. Random lines (approximately 100 S2) developed from TS6 C$_2$ (stress-tolerant population) (Table1), Tuxpeño Sequía C$_0$ (original population) and from Population 21 MRRS (conventionally improved population) were crossed to two conventionally-selected inbred line testers, and topcrosses were evaluated under different degrees of drought stress (Figure 3). Studies were conducted with two other pairs of populations with a common original population but different history of selection for stress tolerance: La Posta Sequía C$_3$ vs Pop. 43 C$_9$, and Pool 26 Sequía C$_3$ vs Pool 26 C$_{23}$. The probability of obtaining a hybrid which will yield 30 or 50% greater than the mean of all hybrids derived from these populations under drought stress was 3-6 times greater when hybrids were derived from stress-tolerant source populations rather than from their conventionally-selected counterparts (Figure 4).

3.2. Line-hybrid relationship under drought stress

Inbred line information indicative of hybrid performance is desirable to potentially reduce hybrid trial evaluation. Development of stress tolerant hybrids requires selection for

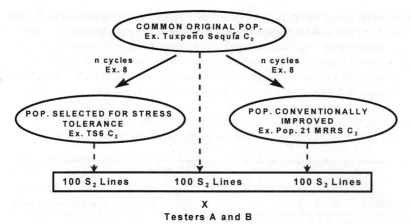

Figure 3 *Development of random S₂ lines from Tuxpeño Sequía C₀, TS6 C₂ and Pop.21 MRRS C₂*

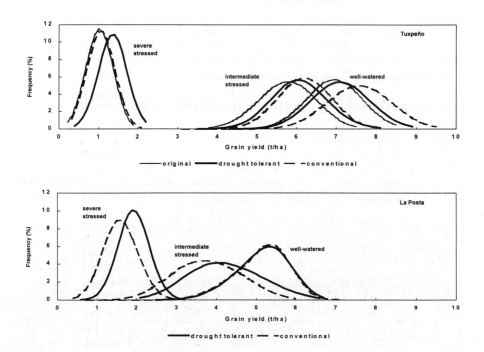

Figure 4 *Frequency distributions of grain yield from different versions of the populations Tuxpeño (top) and La Posta (bottom) in drought-stressed and well-watered environments*

certain traits during inbred line development. In order for this selection to be effective we need to know the relationship between inbred lines and hybrids under drought stress. Betrán et al.[11] examined the relationship between lines developed from drought-tolerant populations and their topcrosses under normal and stressed conditions. One hundred S3 lines derived from TS6 C_1 and 100 LPS C_3 and their topcrosses with Tuxpeño (Ac9021) and non-Tuxpeño (Ac9032) testers were evaluated under SS, IS and WW. Line traits under SS were more strongly correlated with topcross performance under SS than were line traits under normal conditions (Table 3). Selection for a reduction in ASI, SEN and barrenness in the lines under SS could be used to select drought tolerant hybrids.

Table 3 *Phenotypic correlation between topcross grain yield under SS, IS and WW and line traits under SS and WW*

		Topcross GY under SS	Topcross AGY[b] under SS	Topcross GY under IS	Topcross GY under WW
Line traits under SS[a]	GY	0.43**	0.17*	0.05	- 0.22**
	AGY	0.25**	0.07	0.04	- 0.16*
	EPP	0.36**	0.17*	0.03	- 0.15
	AD	- 0.44**	- 0.16*	0.00	0.23**
	ASI	- 0.22**	- 0.18**	0.11	0.07
	SEN	0.22**	0.12	- 0.27**	- 0.30**
Line traits under WW	GY	0.06	0.03	0.20**	0.19**
	EPP	- 0.02	0.05	0.05	0.10
	AD	- 0.51**	- 0.19**	0.08	0.35**
	ASI	- 0.01	0.02	0.20**	0.09

a AD: anthesis date, DS: silking date , ASI: DS-DA, EPP: number of ears per plant, GY: grain yield expressed at 0 g H_2O kg[-1] obtained from shelled grain (SS/IS) or assuming a shelling percentage of 80% (WW), SEN: rate of leaf senescence expressed as the proportion of total leaf area which is dead.

b In order to remove possible escapes due to earliness adjusted grain yield (AGY) was calculated for lines and topcrosses under SS in the following way: $AGY_i = Y_{GY} + R_{GY/AD}$ where, Y_{GY} is the overall mean for GY and $R_{GY/AD}$ i is the residual for each entry coming from the regression of GY on AD.

3.3. Dosage effects and gene action

In CIMMYT's transition from populations to hybrids several issues arise: what is the gene action and dosage effects for DT in hybrids? Do we need DT in both parental lines to get a DT hybrid? What kind of tester is the most appropriate for DT hybrid development?

The first cycle of advanced lines coming from DT populations has been selected and in order to address these issues a diallel study involving five lines developed from La Posta Sequía, four lines from Tuxpeño Sequía and seven elite lines from the lowland tropical subprogram was conducted. Lines per se and their hybrids were evaluated separately in trials planted side-by-side using alpha (0,1) lattice designs. Evaluations are currently in eleven environments: four drought stressed, one low N stressed, and six well-watered environments. Traits measured include grain yield (expressed at 15.5 g H_2O kg[-1]), ears per plant, days to 50% anthesis and silking, ASI, plant and ear height, root and stalk lodging, root capacitance, erect leaves, tassel size and disease scores. Under drought stress, scores were taken for leaf rolling, leaf senescence, and tassel blasting. Preliminary results of grain yield and ASI at four environments, SS and IS at Tlaltizapan (TLSS and TLIS) and two WW environments at Tlaltizapan and Poza Rica (19°N, 60 m elevation)

(TLWW and PRWW) are presented here. Individual analyses of variance were conducted with the SAS MIXED procedure, and a Griffing Method IV Diallel analysis[12] was used to estimate GCA for the lines in all environments. GCA and SCA variance components were calculated using a random model for the diallel design.

Lines were more affected than hybrids under severe stress. Grain yields for hybrids under SS and IS were 13% and 50% of GY under WW. Lines GYs under SS and IS were 5% and 48% of GY under WW. ASI environment means ranged from essentially zero in WW conditions to 8.2 for hybrids under SS. Lines showed shorter ASI under drought stress than hybrids. With increasing drought stress phenotypic variation decreased for GY but increased for ASI (Figure 5). The GCA and SCA genetic variance components for GY were smaller for stressed environments than for well-watered environments (Figure 6). The relative importance of GCA vs SCA, expressed as the ratio between their genetic variance components, increased with stress level (0.70 for WW vs. 2.58 for SS), suggesting the presence of dosage effects and the need for drought tolerance in both parental lines to obtain acceptable hybrid performance under drought. Several lines derived from LPS and two conventionally-selected elite inbred lines, CML 254 and CML 258, had positive GCA values under all environments tested. The line LAPOSTA SEQC3-H297-2-1-1-1-3-# had the highest GCA in all environments.

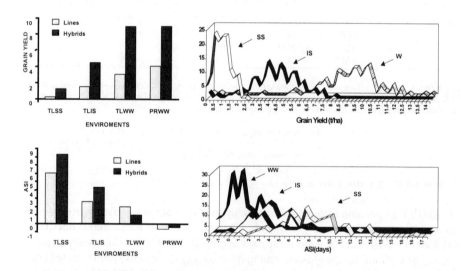

Figure 5 *Means for lines and hybrids (left) and frequency distribution for grain yield and ASI under SS, IS and WW for hybrids (right)*

4. USE OF MOLECULAR MARKERS IN SELECTION FOR DT

4.1. Anthesis silking interval (ASI) QTL mapping and marker-assisted selection (MAS) for drought tolerance.

At CIMMYT, molecular markers (MM) are being used to estimate the number and location of genomic regions responsible for DT that provide the basis for a marker-assisted

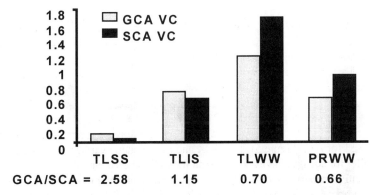

Figure 6 *GCA vs SCA variance components and their ratio under SS, IS and WW in Tlaltizapán and under WW in Poza Rica*

drought tolerance enhancement or improvement of a superior line or population. Ribaut et al.[13, 14] identified and characterized QTL involved in the expression of flowering parameters, with special attention to ASI and yield components under drought conditions. The mapping population was developed from a cross between two lowland tropical lines, P1 (Across 7643) with a short ASI and drought tolerance, and P2 (Across 7729/TZSRW) with a long ASI and drought susceptibility. A total of 260 F_2 individuals were genotyped at 150 loci and their F_3 families evaluated under SS, IS and WW. Two mapping methods were used: SIM[15] (simple interval mapping) and CIM[16] (composite interval mapping). For ASI, seven QTL were identified on chromosomes 1 (two QTL), 2, 5, 6, 8 and 10 using CIM. These QTL accounted together for approximately 50% of the phenotypic variation and were essentially stable over stress levels. For grain yield, five QTL were identified under drought conditions on chromosome 1 (two QTL), 6, 8 and 10. QTLs detected for grain yield and its components were not stable across different water regimes. The comparison between ASI and grain yield QTL showed that four QTL were located at the same position, on chromosomes 1, 6, 8 and 10 (Figure 7). The allele from P1 on chromosome 10 contributed to a decrease in ASI but also contributed to an undesirable reduction in grain yield (with circles in Figure 7).

A marker-facilitated backcrossing scheme using P1 as the donor line and CML247, an elite but drought susceptible tropical line from CIMMYT, as the recurrent line is being conducted to reduce CML247 ASI. QTLs for ASI in this cross were quite consistent with those of the previous mapping population. Using flanking markers for ASI QTL, 8 of 400 BC_1F_1 were selected. In the next backcross a preselection[17] using PCR-based markers was applied to 2300 BC_2F_1 plants. RFLPs will be used for genotyping at further loci of interest, and to evaluate the remaining percentage of donor parent genome in preselected individuals. As a final step, selfed progeny (BC_2F_2) of the selected plants with fixed P1 alleles at QTL of interest will be identified with markers.

4.2. Allelic frequency changes after recurrent selection.

Changes in allelic frequencies in Tuxpeño Sequía during cycles of selection were

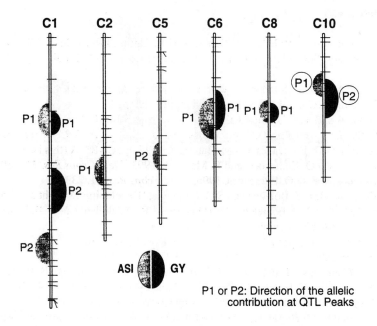

Figure 7 *Location on maize chromosomes of ASI and grain yield QTLs under drought stress detected using CIM for both IS and SS. Genomic regions responsible for the expression of ASI (left) and GY (right) are represented by ellipses for LOD scores higher than 2.0. The width of the ellipse is proportional to the percentage of phenotypic variance explained by that QTL. The parental line contributing the allele for a short ASI or better yield is indicated for each QTL[14]*

estimated using 32 probes chosen based on their hybridization quality, on previous information in the ASI QTL study just described, and on their map position[18]. A total of 116 plants from C_0, C_4 and C_8 were genotyped. The number of alleles identified per locus was between 3 and 7. Most significant changes in allelic frequencies were observed on chromosomes 1, 2 and 6 near the QTL previously detected for ASI. The frequency of the alleles involved in the expression of a short ASI in the F_2 population increases across the cycles of selection. In general, the direction of these changes was consistent across C_0, C_4 and C_8, and their intensity was more marked between C_4 and C_8. If we considered that changes in allelic frequency are mainly due to selection pressure (genetic drift, introgression/contamination and mutation driving forces are not considered), then selection based on the presence of alleles that increased their frequency at specific genomic position across a few cycles of selection could provide a significant improvement of the drought tolerance in the population. MAS on 400 plants from cycles C_0 and C_4 is now underway to evaluate that possibility.

References

1. G.O. Edmeades, J. Bolaños, M. Bänziger, H.R. Lafitte, J.-M. Ribaut, J.W. White, and M.P. Reynolds, Second International Crop Science Cong., Delhi, India, Nov., 17-21, 1996.
2. G.O. Edmeades, J. Bolaños and H.R. Lafitte, 'Proceedings of the 47th Ann. Corn and Sorghum Ind. Res. Conf.', ASTA, Washington, 1992, p. 93.
3. J. Bolaños and G.O. Edmeades, 1993, *Field Crops Res.* **31**, 233.
4. P.F. Byrne, J. Bolaños, G.O. Edmeades, and D.L. Eaton, 1995, *Crop Sci.* **35**, 63.
5. A. Elings, G.O. Edmeades and M. Bänziger, 1997, EUCARPIA (this issue).
6. J. Bolaños, G.O. Edmeades, and L. Martinez, 1993, *Field Crops Res.* **31**, 269.
7. J. Bolaños, and G.O. Edmeades, 1996, *Field Crops Res.* **31**, 233.
8. G.O. Edmeades, J. Bolaños and S.C. Chapman, 'Developing Drought and Low-N Tolerant Maize', G.O. Edmeades, M. Bänziger, H.R., Mickelson and C.B. Peña-Valdivia (editors), CIMMYT, México, 1997, in press.
9. G.O. Edmeades, J. Bolaños, S.C. Chapman, H.R. Lafitte, and M. Bänziger, 1997, Crop Sci., in preparation.
10. G.O. Edmeades, M. Bänziger, M. Cortes, A. Ortega and A. Elings, 1996, 'Developing Drought and Low-N Tolerant Maize', G.O. Edmeades, M. Bänziger, H.R., Mickelson and C.B. Peña-Valdivia (editors), CIMMYT, México, 1997, in press.
11. J. Betrán, M. Bänziger, G. Edmeades, and D. Beck, Agronomy abstracts, ASA, Madison, USA, 1995, p. 88.
12. B. Griffing, 1956, *Aust. J. Biol. Sci.* **9**, 463.
13. J.M. Ribaut, D.A. Hoisington, J.A. Deutsch, C. Jiang, and D. Gonzalez de Leon, 1996, *Theor. Appl. Genet.* **92**, 905.
14. J.M. Ribaut, C. Jiang , D. Gonzalez de Leon, G.O. Edmeades, and D.A. Hoisington, 1997, *Theor. Appl. Genet.*, in press.
15. E.S. Lander, D. Botstein, 1989, *Genetics* **121**, 185.
16. Z.B. Zeng, 1994, *Genetics* **136**,1457.
17. J. M. Ribaut, X. Hu, D. Hoisington, and D. Gonzalez de Leon, 1997, *Plant Molecular Biology Reporter*, in press.
18. J.M. Ribaut, J.M., D. Gonzalez de Leon, G.O. Edmeades, E. Huerta, and D. Hoisington, International Congress on Integrated Studies on Drought Tolerance of Higher Plants, Montpellier, France, Aug. 31 - Sept. 2, 1995.

Breeding Strategies for the Improvement of Nitrogen Efficiency of European Maize Under Low-input Conditions

T. Presterl, G. Seitz, M. Landbeck and H. H. Geiger

UNIVERSITY OF HOHENHEIM, INSTITUTE OF PLANT BREEDING, SEED SCIENCE AND POPULATION GENETICS, D-70593 STUTTGART, GERMANY

1 INTRODUCTION

In most industrialized countries the agricultural production has been dramatically intensified during the past decades due to the increased input of pesticides and fertilizers. This resulted in the pollution of water by nitrate and pesticides in many areas. Therefore, reducing fertilizer input and developing cultivars with improved nitrogen (N) efficiency could be important for future production systems. Field studies indicated, that sufficient genetic variability for N efficiency exists in tropical maize breeding materials and that adaption of maize to low soil N conditions is possible[1-3]. In what follows, we define N efficiency as the ability of a genotype to realize a superior grain yield under low soil N conditions[4]. The following work summarizes the results of a long-term project investigating the necessity and possibility of breeding maize cultivars with tolerance to low nitrogen (N) conditions. The objectives of this study were to quantify the genetic variability of current breeding materials in responding to low N conditions and to assess the genetic potential of landraces as a source of N efficiency. Furthermore, we investigated the usefulness of ear-leaf chlorophyll content and anthesis-to-silking interval as indirect selection criteria for increasing tolerance to low soil N conditions.

2 MATERIALS AND METHODS

From 1989 to 1991, different sets of current European (flint and dent) breeding materials (S_2 and S_4 lines) and landraces (European flint genepool) were evaluated as testcrosses. The lines with the highest yields at the low and high N levels were used for production of 25 'high-input' and 25 'low-input hybrids', respectively. These experimental hybrids were tested in 1992. In 1994, S_2 lines were evaluated for their *per se* performance and for combining ability to two commercial inbred lines as testers. These 48 S_2 lines originated from three open-pollinated landraces which were selected for their performance under low soil N conditions in the testcross series 1990 and 1991. During 1989 to 1991, and in 1994, field experiments were conducted at two N levels (0 and 200 kg N ha^{-1}). In 1992 experiments were carried out at three N levels (0, 100, and 200 kg N ha^{-1}) and at two to four locations in Germany. For each plot the anthesis-to-silking interval, ear-leaf chlorophyll content, grain dry matter content, and grain yield were assessed. More detailed descriptions of the experiments are given by Landbeck[5] and Presterl[6].

3 RESULTS

In order to represent the experiments with testcrosses from 1989 to 1991, the results from 1991 are shown in Table 1. Nitrogen deficiency reduced grain yields by 32 % to 44 % relative to the high N environment (Table 1). In contrast grain dry matter content did not differ between the two N levels (data not shown). Days to anthesis were not much affected by low soil N conditions, while silking was delayed leading to a prolonged anthesis-to-silking interval. Ear-leaf chlorophyll content was reduced at the low N level, as well.

Table 1 *Means, coefficients of genotypic variation (CVg %), coefficients of error variation (CVe%) and heritability estimates for grain yield of 49 flint S_4 lines × dent tester, 49 dent S_4 lines × flint tester and 18 landraces (flint) × dent tester evaluated at two N levels and four locations in 1991*

Statistic	Testcrosses of					
	Flint S_4 lines		Dent S_4 lines		Landraces	
	Low N	High N	Low N	High N	Low N	High N
Mean	4.27	7.62	5.09	7.50	4.04	6.39
CVg %[§]	7.24	5.70	5.03	5.37	8.27	6.40
CVe %	8.03	6.36	8.55	6.87	13.00	8.98
Heritability	0.66	0.72	0.45	0.62	0.64	0.64

[§] All genotypic variances significant at the 0.01 probability level.

Genotypic variances for grain yield were highly significant at both N levels in all sets of material (Table 1). Genotypic variation increased at the low N level in the flint lines and the landraces and decreased in the dent lines. N stress conditions resulted in all sets of breeding materials in enlarged error variances compared to the high N level. Heritability estimates were similar at both N levels for the landraces and reduced at the low N level for the flint and dent lines. A significant increase of the experimental error at the low N level was only observed at locations with heterogeneous soil conditions (data not shown).

Table 2 *Estimated variance components for the performance of 49 flint S_4 lines × dent tester, 49 dent S_4 lines × flint tester and 18 landraces (flint) × dent tester evaluated at two N levels and four locations in 1991*

Source of variation	Grain yield [t ha^{-1}] [§]			Grain dry matter content [%]		
	Flint	Dent	Landrace	Flint	Dent	Landrace
Genotype (G)	12.6**	9.2**	11.6**	3.21**	1.71**	2.47**
G x Location (L)	9.9**	15.4**	15.4**	0.42**	0.47**	0.44**
G x N level (N)	3.3*	4.4*	4.8*	0.01	0.10*	0.02
G x L x N	11.8**	18.2**	9.2**	0.32**	0.36**	0.21*
Error	17.6	22.7	30.3	0.70	0.90	0.98

[§] Components of variance multiplied by 100.

*, ** Significant at the 0.05 and 0.01 probability levels, respectively.

Significant and important variances of interactions between genotypes and N levels for grain yield were found in all three experiments (Table 2). Coefficients of genetic correlation between N levels ranged from 0.7 to 0.8, depending on the materials. For grain dry matter content interaction between genotype and N level was negligible.

In the 1992 trials, the low-input hybrids outyielded the high-input hybrids when treated with 0 and 100 kg N ha^{-1} whereas the high-input hybrids tended to perform better than the low-input hybrids at 200 kg N ha^{-1} (Figure 1).

Average testcross performance of the 18 landraces was rather poor (Table 1). Only three populations were identified with high combining ability to elite inbred lines, particularly under low N conditions. The best S_2 lines derived from these populations were similar or even higher in their testcross yields than an elite check line (Figure 2).

Combined across two locations in 1994, a tight genetic relationship was observed at the low N level between grain yield of testcrosses and anthesis-to-silking interval of lines *per se* (r_g=-1.01). In the same experiment, the coefficient of genetic correlation between grain yield of testcrosses and ear-leaf chlorophyll content of lines was r_g=0.60. Ear-leaf chlorophyll content was also measured in 19 different experiments with the testcrosses of the elite flint and dent lines. Coefficients of phenotypic correlation between ear-leaf chlorophyll content and grain yield under low N varied from r_p=-0.14 to r_p=0.66. For anthesis-to-silking interval coefficients of correlation ranged from r_p=+0.26 to r_p=-0.86 in 28 experiments. Most of the results were non significant and increased N stress did not result in a distinct tendency towards tighter correlations.

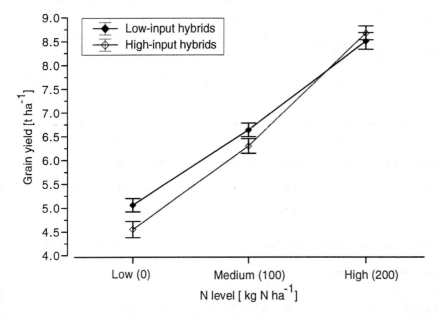

Figure 1 *Response to N supply for grain yield of 25 hybrids selected under high N and 25 hybrids selected under low N conditions, evaluated at four locations in 1992.*

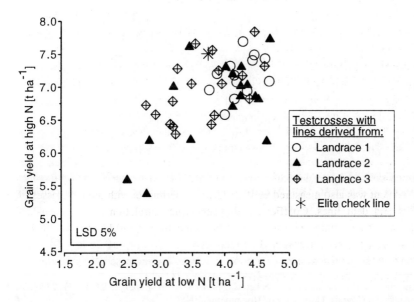

Figure 2 *Testcross means of 48 S_2 lines derived from open-pollinated landraces and one elite check line at the low and high N level, evaluated at three locations in 1994.*

4 DISCUSSION

Environmental means of the two N levels differed clearly, and demonstrated the strong effect of N deficiency. The observed significant and important variances of interactions between genotype and N level for grain yield are in agreement with findings in field experiments with North-American breeding materials[7,8]. This indicates that nitrogen efficiency at low soil N level might not be compatible with responsiveness to high soil N levels. Hence, breeding of specific cultivars that are adapted to low soil N conditions is possible. Expected selection gains were confirmed in the experiments of 1992 where the low- and high-input hybrids were compared. Similarly, Muruli and Paulsen[1] developed synthetic varieties derived from a tropical population at levels of 0 and 200 kg N ha⁻¹. The nitrogen-efficient synthetic outyielded the high-input synthetic and the original population at low N rates, but at the high N level the high-input synthetic yielded the most.

Based on the estimated population parameters, the optimal number of low and high N environments in a breeding programme was calculated assuming different economic weights for the yield under low N. Results indicated that the inclusion of one or two low N sites in a multi-location testing system will increase the total selection gain across N levels without loosing much of the gain at high N[5].

High genetic variability for response to low N conditions was observed in current elite breeding materials. From a breeder's point of view, it is therefore not necessary to go back to landrace materials in order to increase nitrogen efficiency. But the good performance of the tested lines derived from open-pollinated flint landraces indicate the high potential of such materials for increasing the variability of breeding populations.

In experiments with tropical maize materials, Lafitte and Edmeades[2] found moderate to tight and significant phenotypic correlations between grain yield at low N and the indirect traits ear-leaf chlorophyll content and anthesis-to-silking interval, respectively. However, genetic correlations to ear-leaf chlorophyll content were not significant. Results of the present study indicate that these traits are not suitable for predicting the adaptability of current European breeding materials to low soil N, because the respective coefficients of correlation varied extremely across environments and materials.

In conclusion, breeding for adaption to low soil N conditions is feasible and could be done most efficiently by selecting for grain yield in N deficient environments.

ACKNOWLEDGEMENTS - We thank the Ministry of Agriculture of Baden-Württemberg and the Kleinwanzlebener Saatzucht AG (KWS), Einbeck, for financial support as well as for providing genetic materials and field testing capacity.

References

1. B. J. Muruli and G.M. Paulsen, *Maydica*, 1981, **26**, 63-73.
2. H. R. Lafitte and G.O. Edmeades, *Field Crop Res.*, 1994a, **39**, 1-14.
3. H.R. Lafitte and G.O. Edmeades, *Field Crop Res.*, 1994b, **39**, 15-25.
4. B. Sattelmacher, W.J. Horst, and H.C. Becker, *Z. Pflanzenernähr. Bodenk.*, 1994, **157**, 215-224.
5. M. Landbeck, PhD Thesis, University of Hohenheim, 1995.
6. T. Presterl, PhD Thesis, University of Hohenheim, 1997.
7. L.G. Balko and W.A. Russell, *Maydica*, 1980, **25**, 65-79.
8. E.L. Brun and J.W. Dudley, *Crop Sci.*, 1989, **29**, 565-569.

Basis of Maize Breeding for Stress Tolerant and Low Input Demands

L. Pinter[*]

DEPARTMENT OF PLANT AND SOIL SCIENCE, TEXAS TECH UNIVERSITY, LUBBOCK, BOX 42122, TX79409-2122, USA

1 INTRODUCTION

This paper is dealing with the drought as the most serious stress for maize production in continental area, and the genotype dependent N-utilization and shade sensitivity which cause low input demands. The traditional breeding methods are not suitable for improving above traits. New methods that are based on physiological and biochemical traits that control stress tolerance and low input are required. This paper tries to discuss these traits focusing on genotype variability.

2 DROUGHT TOLERANCE

Since the glasshouse effect increases year by year, our climate becomes warmer and dryer, which decreases maize yield or yield stability. Irrigation is expensive in many countries (e.g. in Hungary), that is why the the best way to avoid this stress is the improvement of drought tolerance of hybrids by breeding. In arid and semi-arid areas where every season is dry, the traditionally used selections based on the yield responses are useful. On temporarily drought tolerant territory, like in Hungary, and in many corn-producer countries the traditional method can not be used (*Jonhson and Geadelmann, 1989*). In these areas where water supply changes during growing season, the response to drought stress of hybrids with different length of season can not be compared. However, the water shortage has direct effect on the grain sink capacity such as on the seed set (*Herrero and Jonhson, 1981*) and on the endosperm cell division (*Ober et al. 1991*). At the same time indirect effect as the translocation of assimilates (*Jungens et al., 1978 and Westgate, 1994*) disturbs the evaluation of drought responses of genotypes. This is why only the yield responses of hybrids with equal sink-source relationships can be compared. If these hybrids without classification for length of growing season or sink-source relationship were selected in changeable climate, the chance for breeding drought tolerant hybrids is low. Therefore, instead of using yield responses many traits were evaluated as the model of drought tolerance.

The first approaches of modeling the drought response were done on seedlings under controlled environment. The effects of the drought were determined visually in Hunter's study (*1936*) while *Heyne and Brunson (1940)* measured the killed tissue and dead seedlings. For testing the drought response of a genotype *Pinter et al.* (*1978, 1979 and 1981), Thakur and Rai (1992),* and *Ibarra-Caballero et al.* (*1988*) used the proline content while *Dallmier and Stewart (1992)* tested proline dehydrogenase activity of the leaf. Cell membrane stability was the selection criterion in the studies of *Trapani and*

---

* Fulbright Visiting Professor

Gentinatta (1984), *Trapani and Motto (1984)*, and *Permachandra et al. (1991)*. The plant hormone quantity, especially the amount of abscisic acid (ABA), is one of the widely investigated traits of drought tolerance that select the different genotypes (*Davies and Manfield, 1983*; *Ristic and Cass, 1992*, and *Fobis- Loisy, 1995*). *Ristic et al. (1991)* found correlation between heat shock protein accumulation and genotype responses to drought. Activity of enzymes such as glutathion reductase, antioxidant enzymes were associated to the genotype dependent drought resistance across genotypes (*Pastori and Trippi 1992*, and *Del Longo et al. 1993*).

Correlation of different traits was investigated. Substantial effort was done on the genetic variation of relationship of ABA concentration and stomata conductance (*Zang and Davis, 1990*; *Tardieu and Davies, 1992*; *Tardieu et al., 1993*, and *Tuberosa et al., 1994*). One of the most complete studies was done by *Premachandra et al. (1989)*, who investigated the correlation of among cell membrane stability, leaf water potential, osmotic potential, stomata resistance, cuticular resistance, excised-leaf water retention capacity, K concentration in leaf tissue, degree of leaf rolling, total plant weight and total root length. *Ludlow (1993)* concluded the key traits for surviving drought stress are enhanced water uptake and high osmotic adjustment (OA). All of the traits listed above and many others strengthen these.

2.1 Growth Chamber Experiment for Drought Test

Enhanced water uptake was characterised by the length of stress period (SP) from -0.05 to -0.17 and -0.25 MPa soil water potential in the pots for two drought treatments. This approach differs from the previously used methods where the measurements and the harvest were done after the certain length of stress period. In our experiment the water stress was fixed initiating on the 9 d old seedlings. This method was applied also at the measurement of osmotic adjustment (OA) and water use efficiency (WUE). (WUE is the representative of yield performance of seedlings.)

Table 1
Effect of drought on osmotic adjustment (OA), length of stress period (SP) and water use efficiency (WUE) in seedling stage

TRAITS (units)	OA (MPa)	SP -0.05 to -0.17 MPa (hours)	SP -0.05 to -0.25 MPa (hours)	WUE (g/g)
cv %	30.7	11.0	9.9	15.1
MOST Sensitive	-0.0188 Hybrid 3	78.0 Hybrid 9	122.7 Hybrid 9	166.8 Hybrid 5
MOST Tolerant	0.2892 Hybrid 9	136.8 Hybrid 4	187.0 Hybrid 4	117.8 Hybrid 1

The objective of this study is to show variability among the hybrids in investigated traits (Table 1). The highest cv % was observed in OA, and in SP and WUE these values

same in every soil-level, but they differed from that of in leaf-N-content. As at drought tolerance it was also concluded that the two traits above are controlled by different gene(s).

To demonstrate the different NH_4-uptake under the Hungarian climate only a preliminary experiment was carried out, where the N-treatment and the proportion of $NO_3:NH_4$ were the same as above, but only two hybrids were investigated. The NH_4 contents of the soil from Hybrid 1 were 30.0 and 23.0 % less than from Hybrid 2 in average of 0-90 cm soil levels under the low and high N-supply, respectively.

3.2 Strategy for Improving the NO_3 and NH_4 Utilization

Different genetically controlled NO_3 and NH_4-utilization was observed under Hungarian/Carpatian Basin conditions. Consequently, it is worthwhile to build up a strategy for improving their utilization. The above-used field trial method/soil NO_3 and NH_4 investigation is very time and labor consuming. It requires large plots with homogenous soil. For these reasons, this method is unsuitable for genetics study or breeding. Therefore, physiological or biochemical traits are required to model the genotype responses.

The second step of the strategy is to check the genotype responses with these traits in the different growing stages. This could be followed by the improvement of a population, as was discussed as a general strategy for drought tolerance. The final subjects are the QTL of different traits and build up a complete gene map for the breeders.

4 PLANT DENSITY OR SHADE RESPONSE

The light is the only input that does not cost any money. Consequently, to increase the shade response of the hybrids is the most economical way of improving production. Since there is no treatment to increase the shade response, the only way to do this is through breeding.

The genotype dependent plant density responses were found by *Stinson and Moss (1960)* and were supported by many investigators including the recent paper (*Pinter et al., 1994*). The limit of traditionally used field trial methods for testing the plant density response are: (a) in case of many genotypes it requires huge area, work, and finally money, (b) since the test is based on population level it can not be used for selection e.g. on the plants of segregated population. That is why the model of plant density response is required. In the first approach it seems to be very easy, because it is well known that plant density response basically depends on the genetically controlled shade response (*Stinson and Moss, 1960; Early et al., 1966; Nosberger, 1971*). Consequently, the shading experiment should be a good model. If this method is used for breeding purposes or for screening the plants of segregated population, the yield response is not the right trait. However, one plant can not be grown up under the shaded and control conditions at the same time.

4.1 Previous Approaches for Modeling the Shading Responses

Neither the chlorophyll A and B contents nor their ratios were suitable to model the shading response as was discussed in the last Congress of Maize and Sorghum Section of Eucarpia (*Pinter, et al., 1993*). Only one hybrid was investigated in the shading experiment where the activity of the photosynthesis was measured (*Reed et al., 1988*). However, the differences are promising in the results from the shaded and unshaded treatments. On the other hand the enzyme(s) or other biochemical traits could be worthwhile to investigate as the model of shading responses.

4.2 Strategy for Improving Light Utilization

The final aim is to give an effective tool to breeders to improving the shade-tolerance of new hybrids. Therefore, the QTL of the traits responsible for light utilization is the long term subject from this the gene map could be prepared. Similarly to the above discussed strategies it requires: (a) to find physiological and biochemical traits to model the genotype response expressed by yield performance, (b) to form a suitable gene pool/RILs for the QTL, and finally (c) to conduct the gene map of the responsible trait(s).

ACKNOWLEDGMENT

The presentation is based on the results of common projects of Pannon University, Keszthely, Hungary and Agricultural Research Institute of Hungarian Academy of Sciences, Martonvasar, Hungary supported by FEFA 4 on 1326 projecy number, and the ETH, Zurich, Switzerland financed by Swiss Natural Fundation and COST 814. That is why the author acknowledges the support of Professor Dr. P. Stamp and his colleagues, Dr. U. Schmidhalter and Dr. Boy Feil, Professor Dr. E. Paldi and his colleague Mr. T. Janda, and Mr. Z. Burucs and Mr. Z. Alfoldi who were Ph. D. students at Pannon University.

References

1. K. G. Alexander, M. H. Miller and E. G. Beauchamp, 1991. J. Plant Nutr. **14**, 31
2. Z. Alfoldi, L. Pinter and F. Boy, 1992. J. Plant Nutr. **15**, 2567
3. F. E. Below and L. E. Gentry, 1987. J. Fert. Issues **4**, 79
4. F. E. Below and L. E. Gentry, 1992. Crop Sci. **32**, 163
5. J. Bolanos and G. O. Edmeades, 1991. Agron. J. **83**, 948
6. V. B. Cardwell, 1982. Agron. J. **74**, 984
7. K. A. Dallmier and C. R. Stewart, 1992. Plant Physiol. **99**, 762
8. M. J. Davies and T. A. Manfield, 1983. in: F. T. Addicott, (ed) Abcisic Acid. Praeger. New York, pp 237
9. O. T. Del Longo, C. A. Gonzalez, G. M. Pastori and V. S. Trippi, 1993. Plant Cell Physiol.**34**,1023
10. E. B. Early, R. J. Miller, G. T. Reichert, G. T. Hageman and R. D. Seif, 1966. Crop Sci. **6**, 1

11. I. Fobis-Loisy, K. Loridon, S.Lobreaux, M. Lebrun and J-F. Briat, 1995. Eur. J. Biochem. **231**, 609

12. L. E. Gentry and F. E. Below, 1993. Crop Sci. **33**, 491

13. R. J. Haynes, 1986. in R. J. Haynes, (ed) Mineral nitrogen in the plant-soil system. Academi Press, Orlando, FL pp 303

14. M. P. Herrero and R. R.Jonhson, 1981. Crop Sci. **21**, 105

15. F. G. Heyne and A. M. Brunson, 1940. J. Am. Soc. Agron. **32**, 803

16. C. F. Higgins, J. Cairney, D. A. Stirling, L. Sutherland and I. Booth, 1987. Trends Bioch. **12**, 339

17. J. W. Hunter, H. H. Laude and A. M. Brunson, 1936. J. Am. Soc. Agron. **29**, 694

18. J. Ibarra-Caballero, C. Villanieva-Verduzco, J. Molina-Galan and E. Sanchez-De-Jimenez, 1988. J. Exp. Botany **39**, 889

19. S. S. Jonhson and J. L. Geadelmann, 1989. Crop Sci. **29**, 558

20. P. E. Jung, L. A. Peterson and L. E. Schrader, 1972. Agron. J. **64**, 668

21. S. K. Jungens, R. R. Jonhson and J. S. Boyer, 1978. Agron. J. **70**, 678

22. C. Lebreton, V. Lazic-Jancic, A. Steed, S. Pekic and S. A. Quarrie, 1995. J. Exp. Botany **46**, 853

23. J. A. Lee and G. R. Stewart, 1978. Adv. Bot. Res. **6**, 1

24. Ludlow, M. M. 1993. in T. J. Marby, T. H. Nguyen, R. A. Dixon and M. S. Bonness, (ed) Biotechnology for ariland plants. The University of Texas at Austin pp 11

25. K. F. McCue and A. D. Hanson, 1990. Trends Biotech. **8**, 358

26. J. Nosberger, 1971. Z. Acker Planzenbau **133**, 215

27. R. Novoa and R. S. Loomis, 1981. Plant and Soil **58**, 177

28. E. S. Ober, T. L. Setter, J. T. Madison, J. F. Thompson and P. S. Shapiro, 1991. Plant Physiol. **97**, 154

29. G. M. Pastori and V. S. Trippi, 1992. Plant Cell Physiol. **33**, 957

30. G. S. Premachandra, H. Saneoka and S. Ogata, 1989. Crop Sci. **29**, 1287

31. G. S. Permachandra, H. Saneoka, M. Kanaya and S. Ogata, 1991. J. Exp. Botany **42**, 167

32. L. Pinter, L. Kalman and G. Palfi, 1978. Maydica **23**, 121

33. L. Pinter, L. Kalman and G. Palfi, 1979. Maydica **24**, 155

34. L. Pinter, G. Palfi and L. Kalman, 1981. Plant Breeding **87**, 260

35. L. Pinter, Z. Burucs, E. Paldi and Zs. Magos, 1993. Proc. 16th Conference of Maize and Sorghum Section of Eucarpia pp. 314

36. L. Pinter, Z. Alfoldi, Z. Burucs and E. Paldi, 1994. Agron. J. **86**, 799

37. A. J. Reed, G. W. Singletary, J. R. Schussler, D. R. Williamson and A. L. Christy, 1988. Crop Sci. **28**, 819

38. Z. Ristic and D. Cass, 1992. Int. J. Plant Sci. **153**, 186

39. Z. Ristic, D. J. Gifford and D. D. Cass, 1991. Plant Physiol. **97**, 143

40. W. A. Russel, 1974. Proc. 29th Ann. Corn and Sorghum Ind. Res. Conf. pp. 81

41. L. E. Schrader, D. Domska, P. E. Jung and L. A. Peterson, 1972. Agron. J. **64**, 690

42. K. D. Smiciklas and F. E. Below, 1992. Crop Sci. **32**, 1220

43. H. T. Stinson and D. N. Moss Jr, 1960. Agron. J. **52**, 482

44. F. Tardieu and W. J. Davies, 1992. Plant Physiol. **98**, 540

45. F. Tardieu, J. Zhang and D. J. G. Gowing, 1993. Plant, Cell and Environ. **16**, 413
46. R. H. Teyker, R. H. Moll and W. A. Jackson, 1989. Crop Sci. **29**, 879
47. P. S. Thakur and V. K. Rai, 1992. Environ. Exp. Botany **22**, 221
48. N. Trapani and E. Gentinatta, 1984. Maydica **24**, 89
49. N. Trapani and M. Motto, 1984. Maydica **24**, 325
50. C. Y. Tsai, I. Dweikat, D. M. Hubert and H. L. Warren, 1992. J. Sci. Food Agric. **58**,1
51. R. Tuberosa, M. C. Sanguineti and P. Landi 1994. Crop Sci. **34**,1557
52. T. Weisler and M. J. Horst, 1992. J. Agron. Crop Sci. **168**, 226
53. T. Weisler and M. J. Horst, 1993. Plant and Soil **151**, 193
54. M. E. Westgate, 1994. Crop Sci. **34**, 76
55. J. Zang and W. J. Davis, 1990. J. Exp. Botany **41**, 1125

High Line Performance *per se* under Low Stress Conditions

C. Ipsilandis[1] and M. Koutsika-Sotiriou[2]

[1] DEPARTMENT OF AGRICULTURE PRODUCTION, TEI LARISSA, LARISSA 41 110, GREECE
[2] DEPARTMENT OF GENETICS AND PLANT BREEDING, ARISTOTELIAN UNIVERSITY OF THESSALONIKI, 540 06 THESSALONIKI, GREECE

1 INTRODUCTION

The increasing cultivation of single cross hybrid varieties in maize has caused a greater emphasis on obtaining high yielding maternal lines. The genetic improvement of lines in maize has been accomplished by two general methods of breeding: pedigree selection and recurrent selection (Hallauer, 1990). The commercial maize breeders have seldom used recurrent selection because of its long-term nature (Bernardo, 1996). For this reason Sprague (1984) suggested a compromise method between conventional pedigree selection and long-term recurrent selection.

In this study the application of a compromised method is attempted. The method is a combination of a selection procedure with progeny testing in the form of half-sib (HS) and S1 families, and the proposed honeycomb design for the evaluation of quantitative traits of widely-spaced single plants, grown under low stress conditions (Fasoulas 1973). In a honeycomb design, the units of evaluation and selection at all stages of the breeding programme are individual plants. The selection, is applied from the F_2 generation of a single cross, a generation which has only seldom been used in recurrent selection studies (Onenanyoli and Fasoulas, 1989). In this experiment the source population was the F_2 generation of the commercial single cross hybrid Pioneer 3183.The objective of this research was to evaluate the yield potential of inbred lines derived from the F_2 maternal population with a one year HS-S1 evaluation, (under high stress for genetic input), grown, during the selection, in a honeycomb design i.e. under low stress for environmental input.

2 MATERIALS AND METHODS

The F_2 generation of Pioneer (PR3183) was chosen for this study. PR 3183 is well adapted and one of the highest yielding commercial maize single cross hybrids in Greece. It was introduced in cultivation at the beginning of the 1980s and was deleted by the Commercial Company Pioneer in 1992. During those years the cultivated area with PR3183 reached 50% of the whole area cultivated with corn in 1983, 60% in 1986 and kept the first place between corn commercial hybrids for more than five years (personal communication with Dr I.Sfakianakis, Director of Cereal Institute of Greece). Thus, this hybrid has been considered to possess an optimum combination of favourable alleles for this region as well as sufficient additive genetic variation in its F_2 generation (Hallauer and Miranda, 1981) to guarantee gain from selection.

2.1 Selection procedure

In short, the following procedure was applied. In 1987 the single cross hybrid PR 3183 produced seed by open pollination. The seed thus obtained was planted in 1988 as starting material and is accordingly indicated as generation CO or F_2. The total number of plants was 1200. The intra-row distance was 1.25m and the inter-row distance 1.08m. Each hill was overplanted and later thinned to one seedling per hill. Thus, the density was 0.8 plants per m^2. In this year 512 S_0-plants belonging to the F_2 population were selected by eye on the basis of their vigour and prolificacy. The upper ear of each plant was selfed and the lower ear was open-pollinated.

In 1989 the S_1 lines from the selfed and the half sib (=HS) family from the open-pollinated ear of each one of 512 plants were evaluated in comparison to the original single cross hybrid PR 3183. The evaluation was made by means of a moving block design (Fasoulas, 1983). The entries were located in such a way that, every S_1 row was adjacent to the corresponding HS family, in the present case 512 pairs (i.e. 1024 rows) consisting of an S_1 line and HS-family. Hybrid PR3183 was the check and sown randomly in 64 single rows. The total number of rows was 1088. The intra-row distance was 40cm and the inter-row distance 1m. The density was 2.75 plants per m^2. The plots consisted of 4m single rows of eleven plants each. From each S_1 line a single plant was selected and selfed. Fifty lines, i.e. 10 percent, were selected according to the relative difference in yield with regard to the corresponding HS-family.

2.2 Yield tests

In 1990, from each of these fifty S_2-lines a few plants were grown in an ear-to-row design at a low plant density. From each S_2-line a single plant was selected, selfed and crossed. Experiments aimed at determining the combining ability of the S_2-, S_3- and S_4 lines were conducted in parallel. Thus the S lines were subjected to a selection procedure for combining ability, which diminished the number of lines participating in the selfing programme. At the end, eighteen lines as S_5 (1993) and S_6 (1994) were evaluated in field trials in a randomized complete block design with four replications. The freely available inbred line B73 was added in each randomized complete block design as check. The experimental plot consisted of two 5m long rows spaced 80 cm apart. All plots contained 50 plants, i.e. 25 plants per row. Thus, the density was 6.25 plants per m^2.

All experimental plants were grown using conventional fertilization practices and weed and pest control in order to promote high productivity. Grain yields from each plant or plot were measured after being adjusted to 15.5% grain moisture.

2.3 Data analyses

For the individual plant, yield distributions in CO(=F_2), S_1 and HS and the tests of normality were determined according to Snedecor and Cochran (1967). The null hypothesis, stating equivalence of the entries was tested for grain yield in yield tests. This was done by means of an analysis of variance at the 0.05 probability level. The least significant difference at the 0.05 probalility was used to compare the entries.

3 RESULTS

Single-plant yield distributions for F_2 (=CO), S_1 and HS- families are presented in Fig.1. The actual departure of the yield distributions from normality were shown by estimates of the coefficients of skewness (g_1) and kurtosis (g_2). Strong positive skewness (a long upper tail) was evident in F_2 population, the S_1 values showed a positive skewness, while the HS-values followed a rather normal distribution. The estimates of kurtosis followed a similar trend.

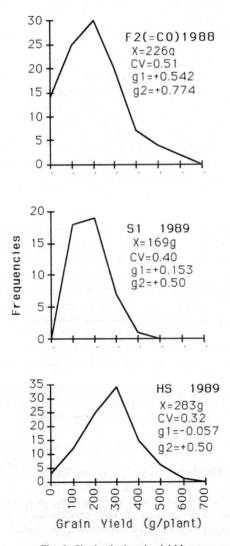

Fig. 1. *Single plant grain yield frequency distributions.*

Table 1   *The mean grain yield performance (Kg/ha) of S5 - and S6 - lines and the yield (in %) relative to that of the inbred line B73.*

S5 - line	Yield	Yield relative to B73	S6 - line	Yield	Yield relative to B73
D - 17	12500	481	D - 17	11000	407
C - 33	10000	385	C - 33	8800	326
C - 35	10000	385	C - 35	8000	296
A - 8	10000	385	C - 22	8000	296
D - 1	9500	365	D - 1	7900	293
A - 29	8500	327	A - 29	7850	291
A - 27	7750	298	A - 27	7150	265
C - 10	7400	285	C - 10	6800	252
D - 8	6000	231	D - 8	5600	207
C - 22	5200	200	A - 8	5400	200
D - 2	4800	185	D - 2	5400	200
D - 5	4750	183	D - 30	5400	200
A - 2	4500	173	D - 5	4250	157
D - 30	4150	160	A - 2	4000	148
B - 24	4100	158	C - 27	3500	130
D - 27	3800	146	B - 24	3350	124
C - 27	3650	140	D - 27	3100	115
B - 36	3200	123	B - 36	3100	115
B73	2600	100	**B73**	2700	100

Total mean ▬ 6600

L.S.D$_{05}$ ▬ ± 2000

cv % ▬ 20

Total mean ▬ 6030

L.S.D$_{05}$ ▬ ± 1600

cv % ▬ 18

Table 2   *The percentage of yield potential of S - families represents the potential of PR 3183.*

	1989	1990	1991	1993	1994
PR3183	100	100	100	100	100
B73	-	17.6	16.5	18.1	18.7
S - Families	32.7	37.3	34.8	46.2	41.9
HS - Families	54.3	-	-	-	-

The results of yield tests conducted in 1993 and 1994 are shown in Table 1. The yield levels varied during the two years of testing, being lower in 1994. However, the productivity of S$_5$ lines or S$_6$ lines exceeded three or four times the productivity of the freely available inbred line B73.

Recent studies have shown that elite inbred lines, which were used for crosses, represented half of a potential single cross (Bernardo, 1996). The present method started with S$_1$ families with a mean grain yield corresponding to 32% of the original hybrid and resulted to 46.2 (S$_5$) or 41.9 (S$_6$) of it (Table 2). The highest-yielding S$_6$ line D-17 represents 76.3% of the PR3183. The yield potential of inbred line B73 kept an unchangable level of yield potential during the breeding programme, corresponding to 16.5 to 18.7 of PR3183 potential.

4 DISCUSSION

The single yield-increasing input that still seems to promise further significant yield increase, is improvement in maize hybrids through breeding (Duvick, 1992). It is well known that the breeding of a maize source population is considered to be an effective way to enhance the development of superior inbred lines (Sprague and Eberhart, 1977), and that proper choice of germplasm determines the ultimate success of selection for genetic improvement (Fountain and Hallauer, 1996).   In this study, the source population PR3183 was the best adapted single cross in Greece.   Furthermore, the quantitative genetic theory (Falconer 1989, p.260) shows that the adaptation to local conditions is due to selection of alleles at many loci for their joint effects of fitness. Therefore, the genetic variability of the F_2 mainly consisted of additive genes.It is reasonable to expect that S_1 progeny performance should reflect mainly additive genetic effects, while HS performance   should reflect some nonadditive effects. (Lonnquist and Lindsey, 1964; Lonnquist and Castro, 1967; Goulas and Lonnquist, 1976). The data showed that: (i) S_1-families have low inbreeding depression, which means high frequency of favourable alleles and (ii) HS-families have low heterosis, which means that the frequency of unfavourable alleles is increased; HS-families with such a genetic backgroud give rise to families with high yield *per se* but with decreased combining ability (Fasoulas, 1988; Koutsika-Sotiriou and Bos, 1996).

Also, in this study a very wide spacing was adapted to enhance differentiation and overcome the disturbing effect of competition among genetically different plants in the segregating generation, i.e. F_2. Additionally, optimal growing conditions were provided to minimize the effects of environmental variation on individual plant yields. Obviously, the favourable environmental conditions were conducive to higher individual plant yields and greater yield differentiation. Selection on a moving block design in S_1- and HS-families and the application of high selection pressure (10%), taken together with the previously mentioned factors, must have positively contributed to the selection response. Results of studies in durum wheat (Michell *et al.*, 1982), in winter rye (Kyriakou & Fasoulas, 1985) and in maize (Onenanyoli & Fasoulas, 1989; Koutsos *et al.* 1992, Koutsika-Sotiriou 1993; Sotiriou *et al.*, 1996) have shown that the selection for individual plant yield is more effective at wider spacing where interplant competition is likely to be less important or absent. Goulas and Lonnquist (1976) showed that combined HS-S_1 evaluation can be effectively employed in selection.

In general, the compromise method combining HS-S_1 evaluation and honeycomb design of sowing, resulted to inbred lines with grain yield potential three to four times more than the elite inbred line B73 or inbred lines corresponding to 70 percent of their initial commercial single cross hybrid.

Summarizing the conclusions of the applied breeding procedure, it becomes obvious that:

(i)   A source population with high yield adaptability has been used;
(ii)   Effects of intergenotypic competition masking the inherent genotypic value are controlled;
(iii)   Differences in phenotypic values are maximized; and
(iv)   Population improvement may be substituted by combined HS-S_1 selection for productivity of lines (*per se*) in low stress conditions during the very early stages.

References

1.   R.Bernardo, *Crop Sci.* 1996, **36**, 867.

2.   D.N.Duvick, *Maydica*, 1992, **37**, 69.
3.   D.S.Falconer, 'Introduction to Quantitative Genetics' Longman Group, London, 1989, p.260.
4.   A.C.Fasoulas, Department of Genetics and Plant Breeding, Aristotelian Univ.of Thessaloniki, Greece. 1973, Publ.No.3.
5.   A.C.Fasoulas, *Euphytica*, 1983, **32**, 939.
6.   A.C.Fasoulas, 'The Honeycomb Methodology of Plant Breeding', Thessaloniki, Greece: A.C.Fasoulas, 1988.
7.   M.O.Fountain and A.R.Hallauer, *Crop Sci.*, 1996, **36**, 26-32.
8.   C.K.Goulas and J.H.Lonnquist, *Crop Sci.*, 1976, **16**, 461.
9.   A.R.Hallauer and J.B.Miranda, 'Quantitative genetics in maize breeding', Iowa State Univ. Press, 1981.
10. A.R.Hallauer, *Maydica*, 1990, **35**,1.
11. M.Koutsika-Sotiriou, *Georgicon for Agriculture*, 1993, **4** (1), 39.
12. M.Koutsika-Sotiriou, and I.Bos, *J.Genet. and Breed.*, 1996, **50**, 7.
13. T.Koutsos, M.Koutsika-Sotiriou and A.C.Fasoulas, *Euphytica*, 1992, **61**, 61.
14. D.T.Kyriakou and A.C.Fasoulas, *Euphytica*, 1985, **34**, 883.
15. J.H.Lonnquist and M.F.Lindsey, *Crop Sci.*, 1964, **4**, 580.
16. J.H.Lonnquist and M.Castro, *Crop Sci.*, 1967, **7**, 361.
17. J.W.Michell, R.J.Baker and D.R.Knott, *Crop Sci.*, 1982, **22**, 840.
18. A.H.Onenanyoli and A.C.Fasoulas, *Euphytica*, 1989, **40**, 43.
19. G.W.Snedecor and W.G.Cochran, 'Statistical Methods', Iowa State Univ. Press, 1967.
20. A.Sotiriou, M.Koutsika-Sotiriou and E.Gouli-Vavdinoudi, *J. of Agricultural Science*, 1996, **127**, 143.
21. G.F.Sprague, and S.A.Eberhart, 'Corn and corn improvement', G.F.Sprague (Ed.). Agronomy, Madison Wi., 1977, pp.305-365.
22. G.F.Sprague, *In*, Proc. 20th Annu.Illinois Corn Breeders School, Univ.of Illinois, Urbana-Champaign, 1984, p.20-31.

Grain Yield Improvement of Elite Maize (*Zea mays* L.) Single Cross Hybrid by Choosing Donors of Favorable Alleles

R. J. Petrović, N. Delić, T.Ćorić, D. Lopandić and K. Konstantinov

MAIZE RESEARCH INSTITUTE, SLOBODANA BAJĆA 1, ZEMUN POLJE 11081 BELGRADE-ZEMUN, YUGOSLAVIA

1 INTRODUCTION

A greater number of groups with divergent material, in which crossing provides the strongest expression of heterosis, is at breeders' disposal. New inbred line development, starting with the best genotype recombination within all groups, is aimed at collecting favorable alleles. Further, breeders have no information whether inbred lines, used for the development of F_2 population, actually differ at loci (i.e. alleles) or not. The results of recombination and new inbred line development from F_2 populations are unpredictable until the very end of breeding procedure. Results of this procedure could be more certain if based on data pointing out if inbred lines crossed to develop the F_2 population differ at particular loci, i.e. alleles. F_2 and BC populations of two related inbred lines are most widely exploited as sources for the inbred line development in maize (*Zea mays* L.), Bauman[1].

The possibility of the target SC hybrid B73 x Mo17 improvement, according to the model of the evaluation of a relative number of favorable alleles[2], is presented in this study. Inbred lines, as potential donors of favorable alleles for grain yield improvement were observed. Three inbred lines as enhancers of grain yield of the target hybrid B73 x Mo17 were selected. In this way, three BC_1 initial populations for selection were developed. According to the pedigree procedure, S_2 families were developed. S_2 test-crosses with an appropriate tester (parent of the target hybrid) were observed trough two experiments. Grain yield, grain moisture content and stalk quality in relation to the check hybrid B73 x Mo17 were the principal criteria for selection of new, improved families in both experiments. Upon the completion of S_2 families testing, self-pollination was continued according to the ear to row method. As a result of this procedure, new, second hand lines SPB4 and MS2, as "improved" versions of the original parental components B73 and Mo17, respectively, were derived. On the basis of the yield higher than in the check genotype B73 x Mo17, the hybrid SPB4 x MS2 was selected in the test trial set up in 1995. The purpose of this study was to evaluate inbred lines as donors for grain yield improvement of the elite SC hybrid and to verify obtained results by the comparison of performances of the "improved" hybrid version with the original hybrid. In this way, the model of the evaluation of a relative number of favorable alleles was tested after Dudley[2].

Among many approaches of clarification of the molecular background of two inbred lines interaction after crossing, dry embryo proteins were, during the last five years intensively analyzed [10]. Results on both the water-soluble and salt soluble proteins in the original and improved maize genotypes are presented in this report.

2 MATERIALS AND METHODS

Inbred lines [3, 2, 7, 9], populations [4, 6] and single cross hybrids could be used as sources of favorable alleles for a particular trait improvement in elite single crosses.

In this paper the following eleven genetically heterogeneous inbred lines, were evaluated as potential donors of favorable alleles for grain yield improvement:

1. L112/8	7. E184
2. L37/1	8. V312
3. V395/31	9. Sp
4. L515	10. S_2
5. ZPE25	11. 172/37
6. L348	

The hybrid B73 x Mo17 was used as a target hybrid for grain yield improvement. All inbred donors were crossed to the parents of the target hybrid, B73 i.e. Mo17. Testcrosses were evaluated for grain yield in a completely randomized block design (RCBD), with four replications at two Yugoslav locations in 1988 and 1989. The target hybrid, B73 x Mo17 was used as a check. Parental inbred lines B73 and Mo17 were grown in a separate but adjacent trial in 12 replications at the same locations. One row plots with 20 plants per plot were used in both trials. Plot size and plant density were 3.32 m² and 62 112 plants ha^{-1}, respectively. Averaged grain yield values over environments for P_1 x D, P_2 x D, P_1 x P_2, P_1 *per se* and P_2 *per se* were used for a relative loci values estimation (μB, μCB, μG):

P_1	parent of the target hybrid *per se* (B73)
P_2	parent 2 of the target hybrid *per se* (Mo17)
D	inbred donor
P_1 x D	F_1 generation of the parent 1 x donor crossing
P_2 x D	F_1 generation of the parent 2 x donor crossing
P_1 x P_2	the target hybrid

The model proposed by D u d l e y [2] was used for a relative loci values estimation This model take includes eight classes of loci for inbred lines, parents of the target hybrid and inbred donors containing favorable alleles for the improvement of certain traits (Tab. 1). The G class locus is the most significant class for the improvement of traits to be dominantly inherited, where the both parents of the target hybrid have negative alleles (unfavorable, recessive alleles), while the potential donor has positive alleles (favorable, dominant alleles). In this way, it is possible to chose the donor (D) with favorable, dominant alleles at loci for which the target hybrid (P_1 x P_2) is homozygous for unfavorable alleles. Symbols A, B, C ... and H represent the number of loci in their respective classes.

The starting point of the applied model for the estimation of loci relative values in this study[2] is: μ is constant for all loci ; the complete dominance a = 1 ; no epistasis; $\mu A = \mu H$.

The genotypic values for the three possible genotypes at one locus are presented such as follows:

Genotype:	Genotypic value:
++	μ
+ -	μa
- -	$-\mu$

a = degree of dominance
μ = one half the difference in the genotypic value between + + and - - genotypes.

Table 1. *Genetic constitution of loci in two homozygous lines and inbred donors*
(after Dudley[2])

LOCI CLASS	PARENT 1	PARENT 2	DONOR
A	+	+	+
B	+	+	-
C	+	-	+
D	+	-	-
E	-	+	+
F	-	+	-
G	-	-	+
H	-	-	-

The frequency of recessive alleles in donor populations at loci C + D and E + F was estimated according to the following mathematical expression:

(I) $(P_1 \times D) - (P_2 \times D) = ((P_1 \times P_2) - P_2)q_j - ((P_1 \times P_2) - P_1)q_k$

q_j = average frequency of recessive alleles at C + D class loci in the donor population of favorable alleles

q_k = average frequency of recessive alleles at E + F class loci in the donor population of favorable alleles

The solution to the equation (I) for q_j, when $q_k = 0$, i.e. q_k, when $q_j = 0$, provides the lower limit frequency of recessive alleles of donors. The solution to the equation (I) for q_j, when $q_k = 1$, i.e. q_k, when $q_j = 1$, provides the upper limit frequency of recessive alleles of donors.

The relative number of loci, where P_1 and P_2 have negative alleles while donor (D) have positive alleles, is designated by μG. The relative number of loci where P_2 and (D) have identical alleles, positive or negative, is designated by $\mu C + \mu F$. The populations with highest estimates of μG parameter, considering the absence of epistasis and with a = 1 (complete dominance), could be used for the improvement of parental inbreds of the elite hybrid $P_1 \times P_2$.

The degree of relatedness between the donor lines and the parents of the elite hybrid was estimated according to the following formula:

(II) $((P_1 \times D) - (P_2 \times D) + (P_1 - P_2)/2)$

The positive value of the expression (II) points to relatedness of the donor with the parent P_1, while the negative value shows the relatedness of the parent P_2 with donor.

The pattern of developing the initial population for selection could be observed by comparing the values of μD or μF (depending on which parent is to be improved, P_1 or P_2) to the value of μG. In the case of the improvement of the parent P_2 there are three possibilities:

1) $\mu F = \mu G$ - the possibility, that the improved inbred shall have more loci with favorable alleles at F and G loci classes than P_2 or the donor, is maximal. Self-pollination of D x P_2 is approached.

2) $\mu F > \mu G$ - points out the back cross (D x P_2) x P_2.

3) $\mu F < \mu G$ - indicates the back cross (D x P_2) x D.

The relative number of loci in classes: B, C, D, E, F and G was estimated according to one of the four models, depending on the frequency of recessive alleles q_j and q_k D u d l e y[2].

Derived families on the basis of results obtained by the relative loci values estimation were tested in S_2 generation with the appropriate tester, B73 or Mo 17. Test-crosses were observed over the two RCBD design experiments with four replications at two locations in 1992. Grain yield, moisture content and stalk quality were the main criteria for families selection. According to the ear to row method, self-pollination was continued in winter nursery. Finally, improved versions of both target hybrid parents were crossed in winter nursery 1994/95 and tested in the trial set up by the RCBD design with three replications at five locations in 1995. The hybrid B73 x Mo17 was used as a check hybrid. Salt-soluble and water-soluble proteins were isolated from dry seed embryo tissue of the original and improved inbred lines. Dry embryos were extracted from the kernel after soaking in distilled water for at least 4 hours at room temperature. Bulk samples consisting of 10 embryos were analyzed.

Reagents:

All chemicals were reagent grade and from various sources.

Protein extraction and electrophoresis methods were used.

The water soluble and salt soluble proteins were extracted after Z i v y et al.[12] and W a n g C h u n et al.[11], respectively. The protein fractionation according to molecular weight was performed by one dimensional polyacrylamid gel electrophoresis - PAGE in the presence of sodium dodecyl sulfate[5].

3 RESULTS AND DISCUSSION

All investigated inbred donors for grain yield improvement of the target hybrid B73 x Mo17 had significant values of the μG parameter (Tab. 2).
Inbreds S_2, L348 and Sp were the best donors. According to the mathematical expression $((P_1 \times D) - (P_2 \times D) + (P_1 - P_2)/2)$, inbred donors Sp and L348 were more genetically

related to the inbred parent B73, while the inbred donor S_2 was more genetically related to the inbred parent Mo17 (significantly negative values of a relationship indicator). In this way, obtained results pointed out that the inbreds L348 and Sp could be grain yield enhancers of B73. S_2 was the most appropriate enhancer of Mo17 grain yield (Tab. 2).

Table 2.   *Relative loci values in hybrid B73 x Mo17 improvement*

DONORS	RELATIONSHIP			CROSSES OF (PxD) TO BE:
	μG	+ = P₁ ; - = P₂	μD or μF	
L112/8	2.35*	0.65*	1.48*	x D
E184	1.47*	-1.42*	1.29	Self
L37/1	1.57*	0.37	1.55	Self
V312	1.00*	1.86*	1.18	Self
V395/31	2.14*	0.14	1.16	Self
Sp	2.49*	1.36*	1.31*	x D
L515	1.07*	2.18*	1.10	Self
S₂	2.92*	-3.23*	0.84*	x D
ZPE25	2.08*	1.20*	1.35*	x D
172/37	1.38*	0.87*	1.43	Self
L348	2.70*	0.19*	1.60*	x D

* > 2SE

The applied model suggests the back-crosses of L348 x B73, Sp x B73 and S_2 x Mo17, with inbred donors L348, Sp and S_2, respectively (μD i.e. μF > μG) in order to develop the initial population for selection.

Considering the results of the two experiments, in which 16 "improved" S_2 versions of Mo17 in the first experiment (Tab. 3), i.e. 9 "improved" versions of B73 in the second experiment (Tab. 4), were observed on the basis of test-crosses; five test-crosses overyielded the check B73 x Mo17. Combination MS₂/18-1 x B73, the third in the range by grain yield values, had a lower grain moisture content than the check B73 x Mo17, while the percent of broken and lodged plants was at the check level (Tab. 3). Also, lower grain moisture content and better stalk quality, compared with the check, are detected in 348B/8-1 x Mo17 and SPB x Mo17 (Tab. 4).

At the very end of a pedigree procedure, two final inbreds MS2 and SPB4, as improved versions of Mo17, i.e. B73, respectively, were crossed. The hybrid SPB4 x MS2 was more yielding than the original hybrid B73 x Mo17 with satisfactory performances of grain moisture content and stalk stability (Tab. 5). In this way, the hypothesis on possibilities for a valid identification of differences at particular loci among related inbred lines used for the pedigree procedure is confirmed.

Table 3. *The performance of the "improved" S_2 families compared to the target*
        hybrid B73 x Mo17 on the basis of test-crosses with B73 in the exp. 1

RANGE	BC$_1$ S$_2$ FAMILIES CROSSED TO B73	GRAIN YIELD (t ha^{-1})	GRAIN MOISTURE CONTENT (%)	LODGED/ BROKEN PLANTS (%)
1	MS$_2$/38-1	9.399*	27.6	7.8
2	MS$_2$/5-1	9.370*	23.6	12.5
3	MS$_2$/18-1	9.343*	24.0	4.3
4	MS$_2$/9-1	9.335*	26.0	1.7
5	MS$_2$/16-1	9.199*	24.2	4.2
6	MS$_2$/35-1	9.048	29.1	7.6
7	MS$_2$/24-1	8.990	26.3	7.8
8	MS$_2$/8-1	8.962	24.8	11.2
9	MS$_2$/3-1	8.877	24.0	4.3
10	MS$_2$/4-1	8.865	25.4	6.5
11	MS$_2$/33-1	8.651	24.6	4.2
12	MS$_2$/7-1	8.553	25.4	7.1
13	MS$_2$/ 2-1	8.499	24.0	20.6
14	MS$_2$/ 10-1	8.339	26.5	1.8
15	MS$_2$/14-1	8.253	24.3	12.2
16	B73 x Mo17	7.826	24.4	4.2
17	MS$_2$/17-1	7.502	26.4	11.2

* $p < 0.05$

Table 4. *The performance of the "improved" families compared to the target*
        hybrid B73 x Mo17, on the basis of test-crosses with Mo17, in the exp. 2

RANGE	BC$_1$ S$_2$ FAMILIES CROSSED TO Mo17	GRAIN YIELD (t ha^{-1})	GRAIN MOISTURE CONTENT (%)	LODGED/ BROKEN PLANTS (%)
1	348B/8-1	10.384*	24.9	6.1
2	SPB/4-1	10.136*	23.3	9.2
3	SPB/37-1	9.601	21.6	9.2
4	SPB/5-1	9.588	24.5	5.0
5	SPB/38-1	9.501	22.9	8.5
6	SPB/12-1	9.092	25.8	5.0
7	348B/12-1	9.055	24.3	8.0
8	B73 x Mo17	8.812	25.9	10.0
9	SPB/29-1	8.318	18.6	3.4
10	348B/5-1	8.251	22.4	9.0

* $p < 0.05$

Table 5.   *Performances of improved and original version of hybrid 73 x Mo17*

Hybrid	Yield (t ha^{-1})	Cob (%)	HOH (%)	Lodg. and brock. pl. (%)
SPB4 x MS2	12.206	17.79	25.25	11.3
B73 x Mo17	11.696	16.67	27.80	13.8

Both the water soluble and salt soluble embryo proteins were isolated from the dry embryo tissue of the original inbreds and their hybrid combination, as well as from improved inbreds and a new hybrid combination. Isolated proteins were separated into numerous components (bands) according to the molecular weight. Electrophoregrams are presented in Figures 1 and 2. Comparing a protein pattern of original parental inbreds and their hybrid combination with the improved combination, both quantitative and qualitative differences are obvious. Heterogeneity among parental inbred lines and the hybrid combination in the original and improved genotypes also existed. Three types of protein bands in hybrid PAGE patterns can be identified as: (1) complementary bands hybrid pattern present two bands, one from the male line and other from the female line; this probably represents the codominance of the parental gene expression in hybrids, (2): bands originated from a female and (3): bands originated from a male parent (for (1) to (3) see Figures 1 and 2). Similar results were reported when different hybrid combinations had been analyzed in the same way [8].

When electrophoregrams of proteins extracted from improved inbred lines and their hybrid combination were compared with those of original parental inbred lines and hybrid combination several new protein bands were identified. This is particularly clear in the salt soluble protein fraction. The next step in the experiments should be characterization of particular proteins and determination of its contribution to both genotype improvement and heterotic effect *per se*.

4 CONCLUSION

Obtained results indicated that salt soluble proteins are a promising tool for the investigation of gene/genotype effect and gene/genotype expression after maize crossing.

Data indicating that two inred lines are different at particular loci are useful in short term breeding programs (pedigree selection). In this way, the possibility for improvement of commercial inbred lines and hybrids is greater and breeding procedure is more certain.

Improved version of target hybrid, SPB4 x MS2, was more yielding and with lower moisture content than target B73 x Mo17.

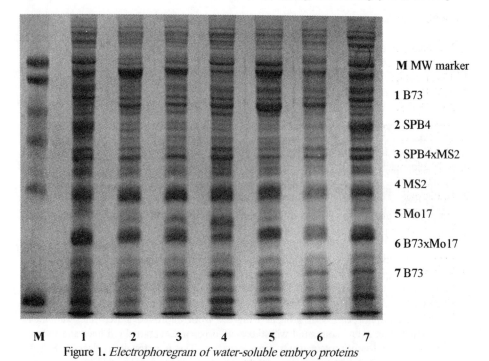

M MW marker

1 B73

2 SPB4

3 SPB4xMS2

4 MS2

5 Mo17

6 B73xMo17

7 B73

Figure 1. *Electrophoregram of water-soluble embryo proteins*

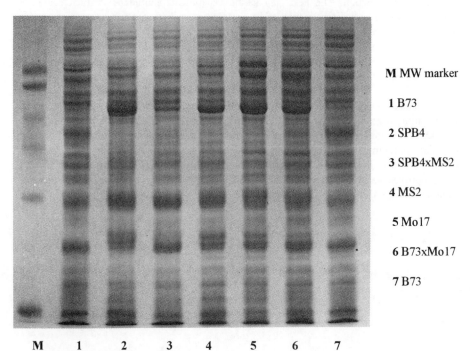

M MW marker

1 B73

2 SPB4

3 SPB4xMS2

4 MS2

5 Mo17

6 B73xMo17

7 B73

Figure 2. *Electrophoregram of salt-soluble embryo proteins*

References

1. L.F. Bauman, Proceedings of the 36th annual corn & sorghum in industry research conference, 1981, p.199
2. J.W. Dudley, *Crop Sci.*, 1987, **27**, p.944
3. J.W. Dudley, *Crop Sci.*, 1984, **24**, p.355
4. J.W. Dudley, *Crop Sci.*, 1988, **28**, p.486
5. U.K. Laemmli, *Nature*, 1970, **227**, p.680
6. D. Mišević, *Crop Sci.*, 1989, **29**, p.1120
7. D. Mišević, *Maydica*, 1990, **35**, p.287
8. S. Mladenović-Drinić, Doktorska disertacija, Univerzitet u Novom Sadu, Poljoprivredni fakultet, 1995.
9. R. Petrović, Advanced Course Maize Breeding, Production, Processing and Marketing in Mediterranean Countries, MAIZE '90. September 17 to October 13 1990, Belgrade, Yugoslavia.
10. A. Tsaftaris, Maize Breeding, Production and Utilisation, 50 Years of Maize Research Institute, Zemun Polje, September 28-29, 1995, Belgrade, Yugoslavia.
11. C. Wang, K. Bian, H. Zhang, Z. Zhou and J. Wang, *Seed Sci & Technology*, 1994, **22**, p.51
12. M. Zivy, H. Thiellement, D. de Vienne and J.P. Hofmann, *Theor. Appl. Genet.*, 1983, **66**, p.1

Subject Index